Lord, Write My Story

Beth (Haggard) Frenzel

ISBN 979-8-89485-109-9 (Paperback)
ISBN 979-8-89485-110-5 (Digital)

Credit for cover design goes to Tamalynn Meyer

All biblical citations were taken from the New International Version of the Holy Bible unless otherwise indicated.

Covenant Books
11661 Hwy 707
Murrells Inlet, SC 29576
www.covenantbooks.com

LORD, WRITE MY STORY

Lord, Write My Story was written as a love letter to my kids, Matt and Ali. Thus, throughout my story, you will see I directly address them.

Commit to the Lord whatever you do, and
he will establish your plans.
—Proverbs 16:3

December 16, 2022

Matt and Ali,

My dear ones, you are my courage to put pen to paper. This is my story, our story, authored by God. He is putting together the jumbled mess so that when we read the whole story, we can see His hand and His presence woven throughout. This was not easy for me to write as I walked through many painful and life-altering memories. For almost a year, I used legitimate reasons that became excuses for why I cannot write this. God has become more insistent as I have struggled to leave the past in the past. I have cried out to God more times than I am proud to admit, asking Him to heal me so I don't let the past ruin my current and future relationships. When I have worn myself out with the crying, I immediately hear Him say, "The healing is in the writing." I realized it's not just my healing, but yours as well.

And maybe others'. I pray that when I go back and read this, I will see that it is His love story for me. I pray I fall in love with the main character, me. All the scripture I've included was given to us and recorded in different notebooks by you and me, Ali. As we prayed over many difficult situations over a period of many years, God spoke commands and promises to us that answered our cries to Him, making it clear so that we knew we were hearing from Him. I compiled them all into one notebook and then decided where they would speak to us in this story. This is written to you, for you. When praying for the title of this book, I kept going to this scripture and imagined Jesus asking me this question. I contemplated what my answer would be. My answer to Jesus's question became the title for this book.

Jesus asked, "What do you want me to do for you?" (Luke 18:41).

I answered, "Lord, write my story."

Though one may be overpowered, two can defend themselves. A cord of three strands is not quickly broken. [Our verse, what ties us together.]
—Ecclesiastes 4:12 NIV

CHAPTER 1

Dare to Dream

In November 2015, I was substitute teaching at Greenfield Intermediate School when a poster in our classroom caught my eye. This was not the first time I had seen this poster. I had seen it countless times before, but that day, I paused and puzzled over the statement Dare to Dream. *Why is it so daring to dream?* I thought. The next day, while on duty at the fire department, I began looking up definitions. I began to think, *What's my dream going to be when I retire? From the time I was five years old, I dreamed—no, I knew—I was going to be a firefighter. From the moment I began to work toward that dream, every goal and every dream rested on the fact that I was a firefighter. But what happens when I retire? Then what? What will my new dream be? Will I have a new dream?* As I contemplated this, I decided to look up definitions, beginning with the word *dream*. One definition led to another, with my thoughts on each word, until Dare to Dream unfolded into my Dream Equation. In this, I connected one definition to another, and it clearly emerged into a statement of fact—a motto, of sorts.

To *dream,* you must have hope. *Hope* is believing something good may happen. Hope means unguarding your heart. *Heart* is the Latin origin for *courage. Courage* is the ability to do something that frightens you and find strength in the face of pain or grief. Courage: I began to think, *How did I find the courage to fight fire or take on the responsibility for someone else's life? Trust:* Trust in my instructors

1

and in my preparation and training. Trust is a firm *belief*. Tracing it backward, the Dream Equation wrote itself.

Belief, trust, courage, heart (now transitioning from my brain and logical thinking), hope, dream.

Belief is the firm foundation upon which trust is built. The strength of courage comes from the heart, which is the compass pointing toward hope. Hope guides us to dream. But first, we must believe!

It is now January 2023, and I'm writing this from your home in Virginia while healing from foot and calf surgery. There have been so many injuries and surgeries over the years. You both have always been my helpers, my stand-ins, and sometimes my caregivers. (Sorry, Matty!) I heal best when I am with you, so once again, here I am. There are so many thoughts bouncing around in my head that I struggle to get started. So I will begin at the beginning.

CHAPTER 2

5-Year-Old Dreams

In their hearts humans plan their course,
but the Lord establishes their steps.
—Proverbs 16:9 NIV

I have spotty memories of my younger years. My earliest memories are from our house on Ritter Avenue in Indianapolis. As the only (or very few) white people in our predominantly African American neighborhood, I recall a boy named Gary, who would bully my brother and me. We played outside a lot. My mom told me stories of how I would sleepwalk, and they'd have to lock doors to keep me from wandering outside. (Times sure were different then!) We moved to the east side of Indianapolis when I was four or five years old. This is where I attended kindergarten. Mom's sister and her family lived one street over. This was a fun time of my life, as I would run through the backyards, cross the street, and play with my cousin. This was also when I began watching a TV show that captivated and fascinated me.

"Emergency" is credited with initiating the desire to become firefighters and paramedics for so many in my generation. I recall grabbing my fire helmet—where I got it, I have no recollection—and watching each episode intently with it on my head. I was wide-eyed, in awe, and certain beyond a shadow of a doubt that *this* was what I was going to do when I grew up! I just *knew* it! It was 1975, and

I was too young to understand that this was not a career that girls entered. When asked what I wanted to do when I grew up, I would emphatically and excitedly respond, "I'm going to be a *firefighter*!" All too often the response would be to ask me again, but in a tone (or maybe even the actual words) that insinuated, I didn't give them an honest response. Not until years later did I understand that they were telling me that it wasn't a job I could or should do. For years, even as the show didn't air anymore, my strong desire and answer did not waver. When I watched a firetruck go by, with its lights on and siren blaring, I knew with everything in me that *that* was what I would do.

The years passed and my dreams faded, becoming a distant thought as my focus turned toward music, softball, and church youth groups. High school became the best years of my life, filled with laughter, choir, and show choir. In my senior year, I was given the privilege to be the middle school sixth grade choir director. I had decided that becoming a music teacher made sense. Plans were made after high school graduation to go to Ball State University as a music education major.

While I no longer answered the question of what I was going to do as younger Beth did, every now and then I would feel that childhood desire stir in me when I'd see a firetruck go by. Somewhere along the way, my interests changed, but I had also begun to believe people who told me girls didn't fight fire.

Dreams and plans for our futures are formed so early, and more often than not, they seem to just disappear one day. Not until I was substitute teaching, one of my many part-time jobs, did I begin to see a pattern of dropped dreams in fifth graders. For so many reasons, I am beyond thankful God overrode my initial belief that I had no business subbing for special education classes as I didn't have any qualifications to do so. Not only because of the amazing and beautiful ladies I had the privilege of teaching with, but I learned so much more from the kids than I ever could have passed on to them. My time in the elementary classroom found me spending a large portion of my days specifically working with Jack and Sally (I've changed their names).

As I began to take them into the intermediate school to socialize and integrate with their peers, I had the opportunity to see the same kids often enough to have struck up conversations about what they wanted to do when they grew up. I always encouraged them and told them to never give up and work toward that dream. As a substitute teacher, I had the freedom to choose which classes I would work in and decided to transition into intermediate school and then junior high with Jack and Sally. By fifth, sixth, and seventh grade, when I would see the same kids I had had the "what do you want to be when you grow up" conversations with, I would ask them if they were still going to do that, and every single one of them quickly and negatively responded with a "No!" I recall that with every response, I would be surprised and say, "Oh! Then what are you going to be?" Most didn't know. It wasn't a dream replaced. It was the death of a dream.

This greatly troubled me. I began to ask myself, *Why did they give up on their dream? When?* To respond that negatively, they had to have been told by an older kid, or, sadly and more likely, an adult, that they couldn't do that and to come up with something different. This breaks my heart! Who has the authority to say what someone is destined to do? So not every little boy with a dream to play in the NBA will make it, but someone will! Maybe the girl dreaming of being an astronaut won't make it the whole way. But *what if*, on the way to attaining her dream, she falls in love with science and realizes she has a knack for it, decides to become a researcher, and discovers the cure for cancer? If she had believed any of the people telling her she'd never make it, humanity would've missed out on her cure. Matt, people will tell you that you'll never become a national news anchor or the president of the United States. Ali, people will tell you that gathering intel on terror groups isn't a job for a female. But someone will get those jobs, so why not *you*?

My thoughts:

- Support dreams. Encourage people. Too many people gave up because someone told them it's not possible. And humanity loses when dreams are discarded.

- No human being has the authority to tell you that your dream isn't meant for you. If God gave it to you, then it is meant for you!
- Stepping into your dream takes time. I was *five years old* when I knew my destiny; *eighteen years old* when I joined the fire service; *twenty-two years old* when I became a paramedic; just turned *twenty-three years old* when I began my fire service career; *thirty-four years old* when, because I followed my dream, I saved your life, Matt. *twenty-nine years* after I believed and embraced my dream! For twenty-nine years, I believed, built, developed, focused, and strove to be my very best. Twenty-nine years, and my dream, fully realized, saved your life, Matt, when you were *five years old*. Never quit! Never settle! Never listen to anyone who tells you it's not meant for you! Someone depends on you to believe and follow your heart.

Jesus replied, "You do not realize now what I am
    doing, but later you will understand."
                                  —John 13:7 NIV

CHAPTER 3

Growing Up

But by the grace of God I am what I am, and
his grace to me was not without effect.
—1 Corinthians 15:10 NIV

While my dream was to be a firefighter, my growing up years were filled with church, family, summer softball, and watching my brother Donovan's sporting events. When I look back, I consider my childhood happy and my high school years the best years of my life— carefree, innocent, filled with laughter, and happy. That's not to say my childhood was all sunshine and rainbows. I am the daughter of an alcoholic dad who just could not overcome his addiction. My dad was a lovable, loving, bright-eyed, wonderful soul of a man who made a friend out of every stranger and would do anything for anybody. He was a Greyhound bus driver and would drink responsibly around his work schedule. Because of his schedule, we didn't see him much. Mom, Donovan, and I were very involved in church, going twice on Sundays and every Wednesday night to youth functions, and I spent my middle and high school years on the Bible quiz team. This was the firm foundation that carried us through not only the coming tragedy but all the struggles that followed. We spent quite a bit of time with family, sharing many meals, sleepovers, and laughs. That's who we were with when we received devastating news that became the first life marker for me. It marked my life before and after, when nothing would ever be the same.

On Labor Day weekend 1984, Mom, myself, Donovan, my cousin, and her family were traveling to southern Indiana to visit and stay with extended family. Dad stayed behind due to work difficulties, but then decided to visit his family in southeast Indiana. It is my understanding that one of his brothers owned a bar at the time, and when Dad got there, he was drunk. When he drove away from there, he wrecked his car on a country road, becoming entrapped. The car caught fire. Dad was either unconscious or trapped and unable to get out; which one, I do not know. It resulted in his death and his being burned beyond recognition. Years later, after I had become a firefighter and EMT, I found his death certificate, which listed the cause of death as "asphyxiation." That told me he didn't die from the impact or his injuries, but was alive as the car burned. Throughout my years in the fire service, I saw my share of burned patients and the agony they endured. My sincere hope is that Dad was unconscious after impact and did not suffer being burned alive.

Because our immediate family was with us, it took a couple of days to find us for notification. Through the questioning of friends, law enforcement was able to connect with someone at our church who knew how to contact the church of the family we were staying with. I will never forget the loud cry of anguish I heard from my mom in a different room of the house. I was troubled and confused. And then I became scared when Donovan was asked into the room with Mom. My thirteen-year-old mind was trying to figure out what would make Mom cry and then ask to see Donovan. I soon found out and wished I didn't have my answer. As I walked into the room and saw Mom crying, Donovan with his head hung, my mom's cousin, and his pastor with sad faces, I looked back at Mom and felt such fear. I had no idea what I was about to hear, but I knew it was going to change my life.

Years later, when I joined the fire department, I immediately and always became the one to not only deliver heart-wrenching news to people that their loved one had died, but my heart of compassion for their anguish gave me the courage to hug them, hold them, cry with them, listen to them, and encourage them. Countless times, I would stare at the deceased patient and be filled with sadness and an unreal knowledge that I knew something that loved ones didn't know yet—knowledge that was about to alter and change their lives forever. And at the very moment that their lives were forever changed, they were still laughing, living, and planning tomorrow with this loved one. It was always humbling knowledge. I'm grateful that God used my childhood heartbreak as deep empathy to comfort and impact many loved ones.

We immediately packed up, piled into the van, and headed home. I hated that ride home. It was suffocating, and I felt trapped. I couldn't escape the crying, low talking, pity looks, and my own chaotic mind. I don't recall arriving home. The coming days rolled into each other with endless knocks on the door and crying. I stayed in my room. I'm thankful this was before instant technological connections, and I couldn't see articles about the accident with pictures. I began to cringe with each door knock because it meant a new round of crying. I plugged my ears. Because it was taking longer than usual to identify him, Mom asked if we wanted to go to school. We said yes. I truly can't say which was worse: constant crying and low talking at home or the silent treatment we got on the bus and at school. At the time, only a couple students had experienced the loss of a parent, and kids didn't know what to do or say, so they would stop talking, look away, steal sidelong glances, and then walk away. I can still recall the feelings of loneliness, intense sadness, and fear. God graced me with the gift of friendship with Beth, who two years prior had lost her mom to cancer and endured similar treatment. My heart had broken at the hurt she carried, and God moved me and gave me the courage to befriend her and walk through her grief with her. In turn, she was my rock.

There is so much my brain chose not to remember during this time. I recall Mom going to Ohio to see Dad's car, meeting my dad's relatives from Kentucky, and seeing so many people throughout the viewing and funeral. The casket was closed, and it was during the viewing that I began to think, *What if it's not him? What if someone else was in his car? Why would he do that? It has to be because he works for the government and had to fake his death. But he's watching us.* That gave me the only glimmer of hope in this entire tidal wave that had swept us away.

My young brain just couldn't accept that he had gone. So often, I felt like an outside observer watching my life unfold, but I didn't feel like it was mine. I knew this wasn't normal, so then I would brighten up and think that I didn't want Dad to see me sad because I just knew he was watching.

I pushed to do my best—always better—to make him proud. I clung to this fantasy until my freshman year of college. It was the same time I found out Mom had not been truthful with Donovan and me about the circumstances surrounding the accident. It was very difficult to grasp that, all these years, I didn't know the truth surrounding the accident. I was furious, hurt, embarrassed, and humiliated. Who else knew the truth? How many people looked at me with pity when I explained how my dad died, knowing I didn't know the truth? Nothing more was ever said about it, and life moved on. No longer did I pretend Dad was alive.

I began to work through two dominant and opposing emotions: anger and sadness. It would always start with sadness at what he had missed—my show choir performances; directing the sixth grade choir; becoming a volunteer firefighter; becoming a medic; stepping into my dream as a career firefighter; joining Indiana Task Force 1; getting married; but most of all, becoming a mom. You both have so many traits that come from him. How he would've loved you! Following the sadness would be anger! He *chose* to miss out on all these. He *chose* to drink and drive. Over time, I accepted he had an illness, and he never would've wanted to miss my life or my accomplishments. Forgiving him was a journey over many years, but it allowed me to accept him as he was and release all negative emotions connected with that chapter of my life.

One of the few pictures of Dad and me. I really liked that dress.
This was my normal short haircut I sported for years. (Circa 1974)

My thoughts:
Do not let fear of stepping into someone's grief stop you. You
don't have to know what to say or even really know them. Just sitting
with them can be enough.

The Lord is close to the brokenhearted and
saves those who are crushed in spirit.
—Psalms 34:18

CHAPTER 4

The Beginning Years

Lord, you establish peace for us; all that we have
accomplished you have done for us.

—Isaiah 26:12

The thought of writing about my years at the fire department is so overwhelming and daunting that it has taken me months of making excuses to finally build time into each week, follow the nudging of the Lord, and hear His promise that He will weave my jumbled mess of memories into words that make sense. However, when you become obedient and begin on a God-ordained path, Satan turns his focus on you and becomes very crafty at knocking you off your path. For all the time I set aside to write, there were nonstop distractions and interruptions. I began to feel God directing me to go on a trip. Through prayer and discernment, I decided to head to the beach for a week. My soul, body, and mind find rest and heal by being near moving water. Satan attempted to distract me by causing me to doubt, worry about not earning money while spending money to go, and feel guilt for looking out for my needs and doing something for myself. I spent the weeks leading up to my departure praying that God would command His army to hold Satan back and to provide a hedge of protection around me, my family, and our households, as I knew that Satan could use that which is most important to me as a deterrent.

And now, I am writing the rest of my story by the water. It is September 11, 2023. I cannot continue with how I began in the fire service without pausing briefly to reflect on the September 11 attacks and the sacrifices demanded and offered up by others. I was on duty that morning and happened to be watching the news when the first plane flew into the WTC. I was dumbfounded. How could that pilot not see the tower? On a crystal clear day? Then I realized it was done deliberately. As my shift partners and I continued to watch what unfolded on live TV, I realized that our lives were forever changing right in front of our eyes.

Fire departments all over the country immediately began lockdown procedures. We were given directives to not leave the stations unless on a run, and we were not to leave our trucks unmanned on scenes. We watched so many firefighters go into the towers knowing what the weight of that much gear and equipment felt like, yet there was no way to fathom the superhuman strength it would take to carry that weight up so many flights of stairs. As we watched the first tower collapse, I had the heart-wrenching realization that we watched most, if not all, of those firefighters walk to their deaths. I just don't have words for the heartbreak. A different place and scenario, and a sacrifice like that, could've been exacted of me. It was a bleak reminder that every day we went to work did not mean it would be like every other day. I'd never heard of Indiana Task Force 1 (INTF-1) up to this point, but this urban search and rescue team hit the road within hours to help their brothers and sisters search and recover. I was on duty the day they came back home, and we stood on the interstate exit ramp to salute the team as they came back home. It was deeply moving. I had no way of knowing that years later, I would have the distinct and humbling honor to serve with this same team. I was privileged to serve on the team with a number of teammates who responded to the September 11 attacks. Four or five of them have since passed away from illnesses associated with working this incident. We get embroiled and bogged down in the everyday tasks and duties of our jobs, becoming numb to the reality that we could lose our lives while doing the job, believing that it won't ever happen to us. And it's only by God's grace that I was able to live to retire.

Many others did not get that opportunity. So today, I remember that fateful day and the sacrifice that was made. I will never forget.

Giving up my dream of becoming a firefighter didn't stop God from starting me on that path. How and when did I finally step onto the firefighting path?

Through friends my senior year, I met a guy who was a volunteer firefighter. I had mentioned my childhood dream to him, and he urged me to join the fire department as a volunteer. I can't even recall filling out the application or when I was officially brought on. It was June 1989, and I was eighteen years old. Back then, things were much less rigid: drinking beer at the station and then going on runs was pretty normal. Not legal or morally right, but sadly, normal. As a matter of fact, the pop machine in the bay had cans of beer in it. I spent the summer of '89 at the station learning how to drive the fire trucks. I had never even driven a pickup truck, and here I was, eighteen years old, learning how to drive a stick shift in a big truck, pulling out and backing into a bay that two trucks shared. Looking back, I cannot believe I had the courage to learn! It was a culture I had never even heard of, much less been around. Young guys who worked night shifts hung out all day watching "training videos"— porn videos. It was the first time I'd ever seen anything like that, and I couldn't believe that men were sitting around watching them together, as if it were a movie.

I began making ambulance runs and quickly realized I was good at connecting with people and liked taking care of them. I was starting Ball State University in the fall as a music education major and began going home on the weekends to make runs. Somewhere during this time, I came back home regularly to take my second class and first class firefighter certification classes and an air crash rescue class, and I was surprised to learn that I enjoyed emergency medical services (EMS) runs more than fire runs. Don't get me wrong, I loved fighting fires, but I knew I was making a difference with people on EMS runs. I enrolled in an Emergency Medical Technician (EMT) class in Muncie my second semester and carried a full college load during the day and went to EMT class two nights a week. I excelled, reading my book cover to cover, over and over again. Everything totally made sense. I realized I didn't want to be a music teacher and wanted to become a paramedic and pursue a career in that field. I acknowledged I needed as much experience as I could get, so while in EMT class, I began riding out with a local 911 service that was really busy.

I tried to ride out five to six hours every weeknight evening when I wasn't in EMT class, then would go home on weekends to catch runs. After EMT class was completed and I was certified, in addition to continuing to ride out with the 911 service and going home on weekends, I joined a small volunteer fire department near college to stay on station and make runs a couple nights each week.

I had my sights set on going to the Methodist Hospital paramedic program, but at the time, they required at least two years of experience before you could even apply. I didn't have the two years yet, so in my freshman year, I changed my major to nursing in order to get science classes and as much helpful knowledge as possible. In the middle of my sophomore year, there was an article in the school newspaper highlighting a new associate's degree in paramedicine that they were building in partnership with Methodist Hospital and would begin the following fall. I immediately made an appointment with my counselor and told him I was switching to that major and to direct me in whatever classes were necessary. I was already taking the necessary classes, and he realized I would be ready before the program was fully functional. They decided to move me through the program before they were completely done designing it. I applied and interviewed for paramedic school and was admitted to begin in the fall of 1992. I loved every challenging, awesome day of medic class. I finished a close second, behind a girl who was going to medical school after our graduation. Cool fact: I was the only one to walk the stage and accept an associate's degree in Allied Health Science-Emergency Medicine in the winter of 1993. The first student to earn that degree in the state of Indiana from an Indiana school.

Before the end of medic school, I was told that the Greenfield Fire Department was hiring a medic. I applied before I had taken my certification exam. (Another cool fact: very few passed paramedic state exam first time through. I was relieved my obsessive studying and practicing found me doing just that. One of our instructors also took the exam with us, and he was the only other one to pass first time through) The state exam (it hadn't changed to the national registry exam yet) had been done two days before I interviewed for the job. I had been told, unofficially, that I had passed, so I could assure them that I was just waiting for the state to process the exam. After a very small agility test, I was offered the job of firefighter/paramedic. The psych test was mailed to my house, and I took it at home. The test consisted of five hundred questions and basically asked the same ten questions in hundreds of different ways. What's the goal—to trick me up? I went back and compared my answers to similar questions, so they didn't think I was a liar. Did I like fire? Umm…yes. That's why I want to be a firefighter. But if I answer it that way, will they think I want to set fires? I spent hours second-guessing the way I should answer. I felt mentally unstable by the end and not at all confident I would get hired. My pension physical was an experience that seemed straight out of a bad comedy movie.

For reasons I cannot tell in this story, suffice it to say I was traumatized, examined in ways I haven't been since, insulted, financially responsible when I was told it was paid for, paraded around the entire office, and stood in a busy hallway to have a lengthy phone conversation with the psych doctor to ask a few more questions before deeming me stable, all while holding my gown shut in the back so I didn't flash everyone. When I left four hours later, I was pretty sure my chances of ever having a career were over. But God stepped in and made that whole hot mess of an experience reveal me as a shining candidate. My start date was November 25, 1993—Thanksgiving Day. What a start!

In the midst of all I was doing, I began dating your dad. We went to the same high school but didn't really cross paths, even though it was a really small school, until I joined Buck Creek FD. He was already a member, and he and a couple other young guys would hang out at the fire department all the time. He was funny, made me laugh, and had patience while teaching me how to drive stick shift.

We began dating and decided to get married. By this time, I was in medic school, so I was planning our wedding while going to school. Our graduation date was set for early summer 1993, and I had confirmed it wasn't going to change, so we set our wedding date for August 1993. But then rescue weekend (a requirement to graduate) kept getting pushed back, thereby pushing back oral boards and graduation. I began to worry that I wouldn't make graduation because, at that point, all wedding details were cemented and could not change. Oral boards took place the day before our wedding and consisted of being run through three or four different scenarios by our medical director and some other ER doctors. It was very stressful and nerve-racking. It was scheduled to last until late afternoon, and I needed to pick up your dad's tux and a few other things before going home to get ready and get to the church for rehearsal. All I could think of was every wedding detail and all I had to do before the rehearsal, so I volunteered to go into every scenario as soon as it opened up, which everybody else was happy to oblige. This kept me from dreading each station and overthinking my way into failing. I sailed through, made it to the rehearsal on time, and made it home from our honeymoon the evening before graduation. By the end of 1993, I had gotten married, graduated from medic school, set up our home, began my career at the fire department, and graduated from Ball State. I would not wish that kind of stress on anybody. And now it was time to settle in and live my best life!

On fire department training ground (February 18, 2019)

On fire department training ground (October 11, 2019)

On fire department training ground (around 2015)

A Time for Everything

There is a time for everything, and a season
for every activity under the heavens:
- a time to be born and a time to die,
- a time to plant and a time to uproot,
- a time to kill and a time to heal,
- a time to tear down and a time to build,
- a time to weep and a time to laugh,
- a time to mourn and a time to dance,
- a time to scatter stones and a time to gather them,
- a time to embrace and a time to refrain from embracing,
- a time to search and a time to give up,
- a time to keep and a time to throw away,
- a time to tear and a time to mend,
- a time to be silent and a time to speak,
- a time to love and a time to hate,
- a time for war and a time for peace.
—Ecclesiastes 3:1–8 NIV

Just as we cannot hold on to or hold off physical seasons, we experience the same changes in our own lives. Some seasons are short, and some appear as if they will never end. Some are beautiful, and others we beg God to end quickly.

Proverbs 10:5 tells us to "know the importance of the season you are in, and a wise son you will be. But what a waste when an incompetent son sleeps through his day of opportunity." Each season provides an opportunity. We either grow from it or we become bitter, wasting our lives away. We choose. Even the seasons that try us are important.

These verses are a summation of my years in the fire service. What better way to recall and recount highs and lows in my career and our lives than to follow the verses and the opposing sets of time. Join me as I travel through these seasons.

CHAPTER 5

A Time to Be Born

Do you remember that over the years, I would tell you that nothing about you was a mistake or wrong? That you were made in God's image, so how could you be anything but beautiful? I believe that God gave both of you, specifically to me, a wondrous gift. Even though my body was not created to easily carry babies, I loved being able to create you and have you grow inside me. Your incredible, blessed, and enduring friendship began when I was carrying you, Matt. I would pretend you were telling me what to say to Ali, and you would have conversations daily. When you were born, Ali already knew you. She was just getting to see you for the first time. You were best friends before you knew it! Being your mom has been life's biggest challenge, one I have fallen short of and failed at more times than I can count. It has also been my reason for going when I wanted to quit—the gift that deeply gives daily—an honor, my joy, my source of laughter. You have been my chattering companions, my representatives, always by my side, my nurse and caregiver, my shadow, and my whole heart. Life literally isn't as bright when you aren't around. I am deeply grateful that you were both born to me and that you love me as you do. It is only by the grace of God that we have the bond we do. But getting you both to the birth stage was no easy matter.

Ali, I was sick from the second we conceived until over a year later. I never vomited but had terrible diarrhea from day one. Which presented a unique issue when I was on duty. My stomach had the worst timing, and I became intimate with many patients' bathrooms while my partner cared for the patient. Talk about embarrassing and humbling! Pregnancy-induced hypertension developed into pre-eclampsia, causing me to be induced a few days early. I had two solid hours to perfect the push, and even though you were small, you were stuck in my birth canal thanks to my hooked tailbone. When your heart rate dropped, they pulled you out with the ole plunger (a suction cup but looks like a plunger). I cracked my tailbone in the process. Add to that the pain of an episiotomy, and sitting was not my friend. After an infection to the epidural site and mastitis (infection in the breast from a plugged duct), I literally felt like I was falling apart. Being a nocturnal being (of which you never really grew out of), most content to be alone, and having difficulty getting you to latch on for breast feeding unless you were held in the "football" position (the position with the least amount of snuggle or touch), I felt like I was failing you as a mom. Lack of sleep did not help with that. I worked on shift until the day before I had you. Imagine what the patients thought when they saw me walk up! I had postpartum depression, which I did not realize until it was diagnosed at my one-year gynecology checkup. Because of all the mental and physical challenges, I had decided I did not want to have any more kids. However, after a couple years, I began to think about how difficult it would be for you if you didn't have a sibling to walk alongside you if something happened to us. So...

Matt, pregnancy started out much easier with you. Until just a few months in, I developed pneumonia, needing to be hospitalized for four days or so. I was so out of it that I forgot to tell them I was pregnant, and they x-rayed me without protection.

I was just praying not only for all your limbs, fingers, and toes, but that you didn't have any extra!

Then, with fifteen weeks left, I was back in the ER with belly pain. I was admitted for pain control for what they believed to be a kidney stone. When it didn't pass in a couple days, they sent me home, and I was then diagnosed with hydronephrosis of pregnancy. As you grew, you compressed my ureter, so it felt like a kidney stone that didn't pass. Pain—a lot of it! Relief comes with delivery, so I had a long haul ahead of me. I continued working on shift until a couple weeks before the induction date. Then I went to days to wrap up loose ends. I worked the day before you were born. Just a week before I came off shift, we had one of the biggest calls in the history of our fire department. It was a Sunday, and I had been having contractions. My doctor told me to hydrate well and rest. I got the hydration down, then we were toned out for an accident on the interstate. This turned out to be the opposite of rest!

I was on Medic 22 (M22), and the wreck was westbound Interstate 70 (1-70), just a little over a mile west of State Road 9. Dispatch kept updating us with more information as we were en route, and it got worse by the minute. It began to sound like a mock disaster where they love to throw impossible situations at you to see how you handle yourself and to test the system. It escaladed from a motor vehicle crash to a bus (my thoughts: okay, it's summer, so it's an empty school bus) to it's a Greyhound bus (okay, that's way worse; maybe it's mostly empty), to multiple calls confirming the bus is on its side, there is a field fire spreading from the bus, confirmed entrapment and multiple bodies in the field around the bus, and confirmed fatality of the driver of the vehicle that crossed the median. All I could think was, *Stop! Stop talking! I need to think!*

We were the first unit on the scene, and never in my almost seven career years nor in the years after had I ever seen anything like this. There was stopped traffic for miles in both directions, tones would not stop dropping as every possible help was started, bystanders started opening our compartment doors and just pulled equipment out and either walked off with it or just threw it on the ground. There were over forty patients, who had thankfully grouped themselves in three areas: those with minor injuries (we call walking wounded-triage level green) were standing off in the distance in the field. Those with moderate injuries were lying in the grassy area at the edge of the road. This group of people were being helped by bystanders (these were triage level yellow). The most critical were scattered in the field all around the bus, and a few were still trapped in the bus. Thankfully, a semi driver had used his extinguisher, so fire was no longer a hazard.

I was supposed to begin triage; however, it was nearly impossible to tell who was helping and who was a patient, and no one could hear me as I called out instructions. Eventually, the message got across that if they were not patients but did not move across the interstate, they would be put on a backboard, loaded into an ambulance, and taken to the ER. That cleared them out. A trauma ER doctor was the first one on the scene, and he used his phone to call for both Lifeline helicopters to respond, so that was a piece of good news. The doctor and I triaged the critical patients and began treatment. My ambulance was stripped of equipment by bystanders and blocked in, so I didn't transport any patients. We were on the scene from midmorning to after dinner. A small bus took the group of walking wounded to our local ER. The yellow patients went to all the Indianapolis hospitals, and the red patients were flown or grounded to trauma centers. I'm thankful I was hydrated before operating for so long in the hot, bright sun, but being hydrated presented its own challenges. Trash cans in the ambulance can be used for more than trash. Just saying.

At my fire department retirement party, you had the opportunity to meet the only baby I delivered in the field and her mom. No matter how amazing it is to have given birth to both of you, it was beyond beautiful to be on the receiving end of a delivery. A bonus is no pain for me!

M22 was called to assist another fire department with an OB. We were initially given no further information, and it is not wholly unusual to have a call like this. My partner and I were just chatting about nothing pertinent on the way there, so there was no reviewing of OB/delivery equipment prior to arrival. We arrived on the scene at the same time as the BLS ambulance. The EMT, a female friend of mine, got out and hollered that the patient had called her and was in the process of delivering her baby. She was excited and running toward the front because she wanted to get in there first and do the delivery. I ran for the back door and found the patient just inside the door. Now all the other personnel on the scene were men, most of whom were highly uncomfortable and did not want to be part of this process. I recall at one point looking out the back door to see no less than four men standing in the yard by the cornfield with their hands on the cot as if it were going to get away from them. My partner had to bring in the OB kit. The patient was crowning, and in no time I'd delivered a healthy baby girl. Not until we were ready to load on the cot for transport did the patient tell me her husband was still on the phone and had listened to the entire delivery. This was landline days, not cell phone. It was so cool to be part of something so beautiful.

Babies aren't the only thing that is born. An idea, a desire, or a plan can grow inside until it is brought into being by action. My belief that I had a talent to truly help others and make a difference was planted in me in the early months after first joining the fire service; by then, I was probably nineteen years old. My first infant death was a SIDS death, and I was one of the first on the scene. It was a very difficult and, obviously, emotional scene for all involved. While we were not able to resuscitate the baby, the doubt I had about whether I could really do this job disappeared, and what was born was confidence and a belief that I was created for this and that this is what I was going to do.

CHAPTER 6

A Time to Die

In their hearts humans plan their course,
but the Lord establishes their steps.
—Proverbs 16:9 NIV

For the Spirit God gave us does not make us timid,
but gives us power, love and self-discipline.
—2 Timothy 1:7

These verses are packed with wisdom that is important for us to remember. You learned how to be an ultra planner because of my jobs and schedule. You learned from watching me. I never chose to just leave on vacation and wander without any game plan. We always had a destination and used the map to plot the most direct course. Do you remember how we'd use the atlas to keep us occupied for hours on road trips? Educational and made the time speed by like the road signs. I believe we need a plan to know where we are going in order to move. It becomes easy to not consult God or to get bent out of shape when our plans don't go as we thought. We can plan our course, but we do not know how many steps God has ordained. He not only knows the number of our days, but He sets our feet in our steps. It's a little like planning our way through the obstacle course of everyday life, but God decides how we move our feet through it.

The definition of *timidity* is showing a lack of courage or confidence; being easily frightened. If we lack confidence or courage, if we are easily frightened, it is not God's plan, nor is it from Him. It is Satan using our human nature as a tool against us. God tells us His Spirit, part of the Trinity, gives us power, discipline, and love. What good is the power to do the seemingly impossible and the discipline to do the necessary if we don't use them? It's the same with love. God is love. He gave it to us, and with courage and confidence, not being afraid of it, we need to accept it and give it.

Had I lived timidly, I would have never been able to accomplish all I did. I would not have been able to impact so many lives. I would never have had the courage to stand in someone else's grief with them or to believe I had the ability to save someone's life.

Sadly, I faced a lot of death in my time. I never grew numb to it, even if my outer demeanor might've reflected that. I either talked through it with my partners or, most often than not, would go to my locker room and quietly reflect and, lots of times, cry. Whether by design or accident, young, old, ill, or healthy, death is not easy to witness. In general:

- Suicides: I always felt a heaviness in my heart for the total despair that gave the individual the strength to follow through. The high school boy who used the shotgun that his grandfather gave him as a gift to the chest but was thoughtful enough to wait until his family left was on the phone with 911 when he pulled the trigger, so they wouldn't come home to find him; the white-haired grandmother, diagnosed with a terminal disease and didn't want to be a burden to her family, who used a handgun to her head in the kitchen while looking at the family pictures she's gathered there. I was eighteen years old, and this was my first suicide. The veteran who suffered from PTSD who put a gun to his head on Veteran's Day with his family just outside the locked bedroom door; the elderly dentist who couldn't watch his wife of many years suffer from a terminal illness anymore and put a gun to both of their heads; the young man who was going to punish the girl-friend he was fighting with and put a gun to his head; a well-respected attorney who put a gun to his chest at home; the husband who had a fight with his wife because he had too many guns, took a shotgun and marched off into the woods, and then she couldn't find him for hours. After pinging his phone, we found him deep in the woods, where he had been for most of the day: overdoses, carbon monoxide, knives—I don't think there's any means of committing it that I haven't seen.

- I had been on the job for a couple years when I responded as an off-duty medic for a brother and sister that were in a bad crash. It was during the county fair, and they were heavily involved in 4H. Because there was only one ambulance, two medics working on two critical patients, and no helicopter that could fly, I didn't have all the equipment I needed since I had gotten on board a basic life support (BLS) truck and just grabbed a little of what I might need from the other medic before they left the scene. Using skills I hadn't used since practicing on mannequins in medic school, she got the full care she deserved. Unfortunately, they both died.

- Responding off duty again, I arrived prior to the only advanced life support (ALS) truck coming to a triple drowning of three young kids from the same family who drowned when the pontoon they were on capsized due to too much weight on board. Working on three kids with two medics, waiting on the helicopter, and contending with a large, panicked family was very stressful. Kids are always the hardest. For years, if the kids were local, I would go to the showing and/or the funeral. It was closure for me, but it also showed the family that their beloved child mattered to me, and I wanted to show respect.

- Not even a year after our motor vehicle crash (MVC), I was on M22 and called for a serious personal injury accident (PI) on the southern edge of Maxwell Intermediate

School. It was just before school started, and there was a lot of bus and car traffic dropping kids off. We got there to find a mom unconscious in the driver's seat with her arm thrown in front of her daughter's chest. The daughter, a student there, was unconscious. We had to cut them both out, getting the daughter out first. I took care of her, but her injuries were too critical, and she didn't make it. Mom was flown out. Normally, we don't transport a deceased patient that we are not actively working on to the hospital, but there was no way I was going to move her out of our ambulance and into the coroner's vehicle where anybody who might have known her would see her or her body under the sheet. Too traumatizing! We put her in the only available room in the ER, a trauma room, and having almost lost Matt in a crash not even a year before, my heart absolutely broke for this family. She had so much glass in her hair that I couldn't let the family see her that way and run their hands through her hair and feel that. I picked as much glass out as I could, all while praying for this family. I did not know until later that her aunt and uncle were in the fire service, and I knew them both. As a thank you, her aunt made me the orange and black fire blanket that is still a favorite of ours to use.

• The deep grief of a spouse for the loss of one they had been married to for many, many years always broke my heart. I recall an elderly couple who had never spent a night apart for all their years. I can't recall the actual number, but they'd been together for sixty-plus years. He called in the morning when he couldn't wake her up. He stated they had been in bed, but she couldn't sleep, so she went out to the recliner. So he went to sleep in his, right next to hers. He said at some point she felt cold, so he put a blanket on her. But she was still cold to touch, so he kept putting blankets on her and didn't want to disturb her since she was finally sleeping. She had been gone for hours and showed definitive signs of death, so I wasn't supposed to work on her.

But the look of complete agony and devastation on his face broke me, so I decided to work on her for his sake. I asked him to go get clothes for her so he could be out of the room when I explained to my partner why I was going to work on her. Her husband said he was going to pack her slippers and housecoat because he thought she would be comfortable in them when she came back home. Then he asked me if I thought that was a good idea. "Yes. If it provides comfort, I think you should get those." My heart…his love. A heartbreakingly beautiful thing to witness.

- Fire service deaths are moving, even if we do not know the individual. The eye-opening reality of what "could" be is no more humbling and visible than witnessing the ceremony to honor the one for whom the ultimate sacrifice was exacted.

- Doyle was the third lieutenant (Lt.) in my career and the one I was least excited to work with. Initially, he was assigned to my shift because the current chief couldn't stand either of us and believed that by putting us together, we would hate life so much that Doyle would just retire and I would quit. We knew what he was doing, discussed it, and made a conscious choice to stand united and prove him wrong. What started as us smiling, laughing, and choosing to be together when we were together as a shift transitioned into a true friendship of genuine joy and laughter. He was old enough to be my dad, and God blessed our unlikely friendship. His family loved me; Doyle loved you, Ali. He always picked you up, carried you around like you were his granddaughter, and played with you. He had had more than one heart attack and regularly had chest pain, needing to take nitro to relieve it. He was my driver the whole time I was pregnant. Can you imagine what the patients' thoughts were? Here's an older man, unbeknownst to them, popping nitro for chest pain, and this whale of a pregnant woman hauling herself into the ambulance with obviously sturdy handrails, probably thinking we need the use of the ambulance for care more than they did. I will say that we always gave great care and that no one suffered or wasn't properly cared for. We had been partners for a few years when he retired. Not long after, he developed cancer. I was on a task force exercise at Camp Blanding in Florida when Doyle's son called to say that Doyle didn't have much time and he wanted to see me. I was devastated. I told him my situation, and he put Doyle on the phone. We chatted for a few minutes; his voice was so weak I could barely hear him, and when I hung up the phone, I was broken. I knew I had just said goodbye to my friend for the last time. His son called me a couple hours later to tell me that he had died and to ask if I would speak at his funeral. This was one of many examples of God making good what evil intends for bad. God blessed this unlikely partnership, growing it into one of beauty and blessings.

- One final impactful event. INTF-1 had been deployed to North Carolina to assist with Hurricane Florence. In our first mission, I was assigned to the squad going out to assist the local fire department. We didn't have any further information than that, and since the hurricane was still hitting the area, we assumed it was for a water rescue assist. We were not prepared for what we found when we got there. I will never forget the scene that first greeted us. One rescue truck with a crew of men sitting on the back bumper, looking exhausted and completely dejected...defeated. The scene was a house with a ginormous tree lying across the back of it, crushing the entire rear of the house. The scene commander informed us that there were no units to assist because they were across town fighting a couple of structure fires. This crew had been on the scene for hours trying to access the trapped victims. The father had been rescued and transported, but informed them that when the initial storm hit, he and his wife brought their baby boy into bed with them. They had gained a small access point to the mom and knew that she was deceased, but could not find the baby. They needed our help in finding and (hopefully) rescuing the baby, as well as recovering the mom's body. A game plan was put in place, vehicles moved for better access, and we got to work. I was the only medical specialist on scene for the first couple of hours. Our commander called for more squads and equipment. I was so proud to be part of that crew, watching the local firefighters and our members work side by side with no egos, a common goal, and respect for all personnel and their expertise. When enough access was gained to get near the mom, I was called up to climb in and under the tree to see if I could get to the baby. This was the hurricane that didn't move off the coast for days, thereby, picking up the ocean and dumping it inland, wreaking havoc on the entire town. These incredible men had been working tirelessly through multiple rounds of this hurricane dumping, with huge trees bending to the ground and many breaking branches, making the most of every lull in the storm to make headway. In a quiet period of total stillness, I climbed in, felt the mom, and made access to the baby's arm. With years of experience and training that the best in the country could provide, I knew for a few reasons that this was not a rescue but a recovery for the baby boy. I also knew that when I delivered this news, it would devastate the men standing outside the house. I said a prayer for these men, backed out, looked at my rescue team manager, and shook my head no. Every single one of the original crew members hung their heads and, with slumped shoulders, cried. Deeply. And so did we. Even now, I tear up. We got back to work, but without a hurried pace. Deliberately and respectfully, the work continued until we had full access to both the mom and baby. At this point, we stood back and let them remove and place mom and baby in the same body bag and, respectfully, carry them down a line of us blocking the public view. After putting all the equipment back, we stood as a group in the same spot this group of men kneeled and prayed as we were en route to assist them and discussed the scene. Following the scene debrief, with some standing and others kneeling with bowed heads, we prayed. It was one of the most impactful moments of my life. It was captured by the news, reported on by Lester Holt for the *Today Show*, and mentioned by him in an interview later as one of the top ten most memorable incidents of his career. This became something that haunted me, presenting as flashbacks when I least expected it, troubling me and bringing me to tears often. I'm sure a complete lack of sleep and all that we were physically enduring during this long hurricane response made this worse. I second-guessed myself probably a hundred times after I climbed out and shook my head no. What if I was wrong? What if he was alive? What if they took their time because this was not a rescue and he was still barely alive when they gained full access to him? What if I am wrong? I started and stopped myself from walking up to the house to double check, arguing with myself countless times. I *know* what death looks and feels like. I am not wrong! I doubted. I worried. I argued. And I left myself with doubt. Only after I was home, rested, and had quiet time to deeply reflect and get counseling help did I realize why I was so troubled. There was only one previous time in my entire career that I questioned myself, and that was when I was trapped in the car with you, Matt, during our car crash. I debated and argued with myself about the best thing to do to help you, then questioned that decision when I tried to get you out of your car seat and your legs were tangled and stuck. Your limp arms, not that much bigger than this baby boy, my questioning of my decisions, my complete distress in wondering if I had made an error that might negatively affect the outcome—I was superimposing both incidents and escalating my fear with our accident and all that encompassed. All of which I full well knew I had not dealt with but shoved into an imaginary box and neatly packed up and locked away to be dealt with on a future date that was far in the future. I had no idea how to process all we lived through and I didn't have the time to figure it out or get help. Being honest about the hurricane incident meant pulling the box with all the negative feelings of our accident back out, opening it up, and honestly sorting through all the intense facts, hurts, and terrible changes that were made after it. That's a story for another time—a story I will allude to some more in this book. That baby boy, whom I did not get to lay eyes on, never knew his name, and only was able to touch his arm, touched me so fully and deeply that I will never forget him and will always be deeply moved when I think of him and bowing my head in prayer with his heroes.

Hurricane Florence team picture. I'm sitting in the boat with the dog. (September 2018)

Sitting in a military vehicle that transported us to our search areas for Hurricane Florence. Photo credit: Dr. Chris Strachan. (September 2018)

We each have our time to die. Days ordained by God. As we know, Matt, for those chosen, literally only God knows why, to have their days altered or re-ordained, if that is a thing, God then places a mantel of responsibility to tell the story of days gifted back. I will fully tell our story, the events leading up to it, and every detail about our accident after I have written my story. I will need God to write that story as well.

CHAPTER 7

A Time to Plant

Commit to the Lord whatever you do, and he will establish your plans. [Another version says and your plans will succeed.]
—Proverbs 16:3 NIV

Similar to the proverb from the last chapter, we are not lost with no idea where to go if we commit whatever we are doing to the Lord. He will establish your plan and your steps. That doesn't mean that everything we do will succeed by our standards, especially if they are our plans and we didn't pray about them. We commit by praying about it, presenting it to God, submitting our will to His, and giving it to Him. He then lays out our plan, each step of the way. We follow, and we succeed. His timing, not ours. And "succeed" will be for what God wanted to happen, not necessarily what we wanted.

Be still before the Lord and wait patiently
for him; do not fret when people succeed in their
ways, when they carry out their wicked schemes.
(Psalms 37:7 NIV)

Patience is most definitely not a virtue of mine. It hasn't gotten better with age. It has, however, gotten easier to take a deep breath and to allow impatience to shrink. We sure know what waiting is. Through the long years when you were still going to your dad's, we may not have waited well, but I believe the waiting taught us to choose laughter, happiness, togetherness, and love. It was in the years of waiting that our cord of three strands developed and strengthened. I don't think patience comes without stillness. I notice that I become increasingly impatient when I'm super busy and cannot be still. There is value in being still and spending time with God.

> The Lord himself goes before you and will be with you; he will never leave you nor forsake you. Do not be afraid; do not be discouraged. (Deuteronomy 31:8 NIV)

Forsake is not a word we hear much, but it means to abandon or give up. There is so much promise in this verse. God has a plan for you. He establishes your steps, and He will go before you to prepare the way and then be with you. He says he will "never" leave you. He won't. No matter how crazy it seems or how scary, He is with you and will never abandon you. So have no fear, and don't be discouraged if no one else can see or understand your path. God is on it, so you know it's not wrong. You both feel deeply what the word *abandon* means. Matt wrote it clearly on a bookmark he made in crayon one Sunday: God is my Father, and heaven is the best. How profound. Matt, you made that (I still have it!) when things were particularly tough at your dad's. Your dad fell short and eventually decided neither of you were worth keeping in his life. Truly, his complete loss. Your words told me that though you hurt, your true father is God. And you know He will never abandon you.

> So do not fear, for I am with you; do not be dismayed, for I am your God. I will strengthen you and help you; I will uphold you with my righteous right hand. (Isaiah 41:10 NIV)

> For I am the Lord your God who takes hold
> of your right hand and says to you, Do not fear;
> I will help you. (Isaiah 41:13 NIV)

When I look at all these verses together, I see clear commands and promises: Do this or do not do this, and He will do this.

Commands:

- commit
- be still
- wait
- be patient
- do not fret, fear, be discouraged, be afraid, be dismayed

Promises:

- He establishes your plans.
- Your plans will succeed.
- God will go before you.
- God will be with you.
- He will never leave you.
- He will never abandon you.
- He is with you.
- He will strengthen you.
- He will help you.
- He will uphold you.
- He will take hold of your hand.

What does all this have to do with planting? We usually see planting as putting something in the ground, tending to it, watching it grow, harvesting it, then enjoying the fruits of our labor. There is something to be said about growing our own garden food. It tastes better because we put the time and effort into it. Our minds are fertile ground for seeds of doubt, fear, insecurities, distrust, lies, hate, and evil to be planted by Satan. If we do not guard our minds from what Satan planted, we unknowingly water, tend, and help it grow into a field of darkness, and soon we lose our way. We must choose to live and believe in the promises that God gives us. If we fill our minds with goodness, God's word, and His promises, then there isn't room for Satan's seeds to take over. Read God's word, believe it, and trust that it is true.

CHAPTER 8

A Time to Uproot

"For my thoughts are not your thoughts, neither are
your ways my ways," declares the Lord. "As the heavens
are higher than the earth, so are my ways higher than
your ways and my thoughts than your thoughts."
—Isaiah 55:8–9 NIV

My nature is the exact opposite of uprooting. I am a very rooted person, to the point where I am stubborn and maybe a little obstinate.
When I begin doing something, I tend to do it until God removes it
from me. The one area that I can relate to uprooting is when I joined
the task force. Sudden deployments that yanked me out of my normal
life to spend weeks being fluid and going with the flow dictated that
you both were made to do the same by picking up my responsibilities
at home and representing our family in my place. Often times, when
God moves in our lives, we resist because we're fine where we are; we
don't want to move even when we're not fine. The fire department
culture, personnel, and run load were becoming more than I was able
to bear after all those years. While I was living in God's plan for my
life, most of the men I worked with over the years just couldn't accept
that I belonged there; they wouldn't join with my talents, skills, and
capabilities to be better as a whole for those we were sworn to care
for. While I loved what I did, I detested having to do it with people
who hated me. I increasingly cried out to God, saying, *God, please!*

I can't do this anymore! Just give me the strength to make it _____ more years or months. God knew I wouldn't make the decision to uproot myself, so He decided to answer me in His way, in His timing.

His plan to uproot me was in motion years before it happened. A spinal cyst ended my career. My lower back began hurting in mid-2020, long before it was found. I tolerated it, went to the chiropractor, refused to spend the money for a CT scan, and ignored the creeping left lower leg numbness, severe leg pain, drop foot, knee buckling, and thigh cramp symptoms I was having because I just didn't have time to address them. Being deployed to the Surfside building collapse with the task force at the end of June 2021 accelerated the cyst growth so that I barely could make it into the ER on Labor Day Sunday, 2021. A CT scan showed the cyst in my spine was herniating L4, L5, and S1 and had completely compressed the sciatic nerve, which explained the left leg issues, and was compressing the end of the spinal column nerves. I prayed for the right surgeon since working around my spine could've ended badly if not for the right doctor for me.

The doctor I prayerfully chose saw me Tuesday morning, two mornings after my diagnosis. I was so hopeful he could just aspirate the cyst in the office that I took Mom with me to drive me home after the procedure. How wrong I was! He wanted to operate on me two days later, but insurance wouldn't approve it that fast, so it was scheduled for September 21. When he told me what he was doing and how long I would be off, I knew I could never go back to the fire department or the task force and the load my spine would have to carry with both jobs. He really was the best doctor for me because he let me process and come to that conclusion on my own. The surgery took over four hours. I had a four-day hospital stay and was restricted to only being able to stand, sit, or lie down for four months. Two rods, four screws, three spacers, bone grafting in between L4 and L5, removal of the golf ball-sized cyst, removal of quite a bit of bone, and all the supporting structures—an incision that went almost half the length of my spine—brought me to where I am today.

After a year and a half of appointments and x-rays every few months, I was released without restrictions. He had agreed I could continue as a paramedic at the horse track since it's not repetitive heavy lifting. This healing period was one of the toughest months of my life and part of the toughest season I have ever gone through. I am still in that season, coming out of it. Having you both there when I came home from the hospital is a testament to the strength of our cord and meant everything to me. I don't want to minimize the depth of my gratitude for how God directed this season. He graciously showed me how the timing had been His all along. I wanted to get to the first of the next year so I could get the rest of a raise that would boost my pension check and "work" one more shift of the new year so I could grab my paid days off for that year. I had exactly enough paid days to take off, which got me through my last shift of the year. After taking all my days together at the beginning of the next year, I officially retired in mid-March 2022. Due to a couple weeks off, then Surfside deployment, then coming back with COVID-19, off four months for my back, minus a couple shifts I worked in this time, added to the two and a half months off at the beginning of 2022, I was off nine months paid. Such a blessing!

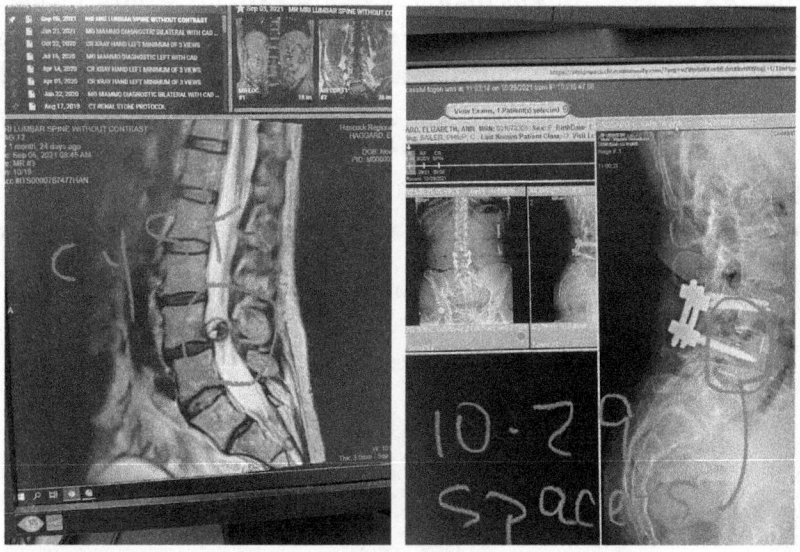

Imaging of spinal cyst and hardware post-surgery (October 29, 2021)

Matt and Ali came home to help me the day I arrived
home post-surgery (September 25, 2021)

God uprooted you when you followed His nudge to move to Virginia. It was most definitely not one of my finer moments as your mom. You came to me to run the thought by me, and I lost it. I opened my mouth, and fear spoke words that I didn't mean. I knew the day would eventually come, but I couldn't imagine life without either of you close by. I'm thankful you forgave my words, included me, listened to my suggestions, and let me be part of the process. You went to a state ten hours away, where you knew no one at the start of the COVID-19 pandemic, no job, and a truckload of faith that if God brought you to it, He would see you through it. And He did, every step of the way. You both, individually and together, have blessed so many people there. I am in awe of you both and so incredibly proud!

CHAPTER 9

A Time to Kill

Do not repay anyone evil for evil. Be careful to do what
is right in the eyes of everyone. If it is possible, as far as it
depends on you, live at peace with everyone. Do not take
revenge, my dear friends, but leave room for God's wrath, for
it is written: "It is mine to avenge; I will repay," says the Lord.
Do not be overcome by evil, but overcome evil with good.
—Romans 12:17–19, 21

This season is a tough one to understand. I think we assume it means to kill a person, especially when its opposing season is to heal. But I think it realistically means to kill unhealthy relationships, bad habits, or anything we do that is unpleasing to God. Matt, I used to talk with you about knowing when it was time to leave your dad's and the hatred and mental and emotional abuse you endured while there. You lived out overcoming evil with good time and again, year after year, but I believed a time was coming that God would avenge, repay them, and unleash His wrath on them, and I did not want you to stay past when God said to leave and get caught in His wrath.

I responded to a number of murder scenes, but there was one that stood out for me. I was on M22, and we were called for an unconscious child. When we arrived, we were told they were bringing the baby out to me. The baby girl was limp and laid on my cot, and we immediately began working on her. I saw and pointed out obvious bruises to the law enforcement officer. As the story unfolded, we found out that the mom's boyfriend was watching her while mom was gone, got mad at her for crying, and, I believe, shook her. He realized what he had done and decided to create an alibi for himself by putting her in her stroller, covering her with a blanket, pulling the shade over her, and walking quite a distance to a gas station, where he went in to purchase something. This was captured by a number of cameras in that area, and at least one woman saw her and believed she was asleep. He did this so she was seen, and he could say she was alive when he took her for a walk. There is evil among us, and we just cannot understand it. We may not see it, but if God says He will avenge, then we must believe He will.

CHAPTER 10

A Time to Heal

I have the most scripture to add here because, man! From injuries to surgeries to the mental and emotional abuse we've all had to endure, I have searched out scriptural hope for healing. I'll lay all the scripture out, then comment.

> Though one may be overpowered, two can defend themselves. A cord of three strands is not quickly broken. (Ecclesiastes 4:12 NIV)

> But those who hope in the Lord will renew their strength. They will soar on wings like eagles; they will run and not grow weary, they will walk and not be faint. (Isaiah 40:31 NIV)

> Peace I leave with you; my peace I give you. I do not give to you as the world gives. Do not let your hearts be troubled and do not be afraid. (John 14:27 NIV)

> Yes, my soul, find rest in God; my hope comes from him. (Psalms 62:5 NIV)

"For my thoughts are not your thoughts, neither are your ways my ways," declares the Lord. "As the heavens are higher than the earth, so are my ways higher than your ways and my thoughts than your thoughts." (Isaiah 55:8–9 NIV)

Why, my soul, are you downcast? Why so disturbed within me? Put your hope in God, for I will yet praise him, my Savior and my God. My soul is downcast within me; therefore I will remember you from the land of the Jordan, the heights of Hermon—from Mount Mizar. Why, my soul, are you downcast? Why so disturbed within me? Put your hope in God, for I will yet praise him, my Savior and my God. (Psalms 42:5–6, 11 NIV)

Wait for the Lord; be strong and take heart and wait for the Lord. (Psalms 27:14 NIV)

The Lord will vindicate me; your love, Lord, endures forever—do not abandon the works of your hands. (Psalms 138:8 NIV)

Look at the nations and watch—and be utterly amazed. For I am going to do something in your days that you would not believe, even if you were told. (Habakkuk 1:5 NIV)

Do not be anxious about anything, but in every situation, by prayer and petition, with thanksgiving, present your requests to God. And the peace of God, which transcends all understanding, will guard your hearts and your minds in Christ Jesus. (Philippians 4:6–7 NIV)

Trust in the Lord with all your heart and lean not on your own understanding; in all your ways submit to him, and he will make your paths straight. Do not be wise in your own eyes; fear the Lord and shun evil. This will bring health to your body and nourishment to your bones. (Proverbs 3:5–8 NIV)

"For I know the plans I have for you," declares the Lord, "plans to prosper you and not to harm you, plans to give you hope and a future. Then you will call on me and come and pray to me, and I will listen to you." (Jeremiah 29:11–13 NIV)

When we moved to Independence Rd, I decided and announced that our home would be our sanctuary, our safe place, where we could be ourselves, feel what we needed to feel, and know that we would always be loved. If you did not support us and love us, then you could not cross our threshold.

We guarded that. I regret that more times than I care to remember, I failed at this. Sheer exhaustion—the weight of what was happening to me at the fire department, navigating the nonstop games your dad was playing, and how to help you—left me weary. I'm ashamed to say that my temper was quick to flare, and I didn't always say words that showed love or built you up. I hurt you, and it is only by the grace of God and your beautiful forgiveness that you love me. Somewhere in the mess of our lives, we declared Ecclesiastes 4:12 as our verse. We began to live out this verse. Over and over again, we found there were always one or two strands that were getting stretched to the max, almost to the breaking point, but the third strand was strong and held the cord strong until we could find relief. Our cord is strong because we believe in it and because God is with us, binding our cord. We pull strength from each other, we allow rest and healing for each other, we discuss big decisions with each other, we pray for each other, and we support each other. When it's time to heal, our strands pull together.

Hope brings healing. The months following back surgery were very bleak. I was deeply troubled and scared. Most people in my life didn't know what I was going through, and I disappeared. I cried hours every single day for months. God would send verses on hope, not as a wishful-thinking kind of hope but as true hope. God met me in those dark days. Even now, I am overwhelmed with gratitude for His clear provision. I wrote down and many times each day read the last six verses above.

I would scoff when I read to "wait and be strong." *Okay! Yeah, I'll wait because there's nothing else I can do. And I'd love to be strong again.*

The Lord will fulfill His purpose for me. For the first time ever, I couldn't make something happen out of sheer will and work ethic. I had absolutely no idea what my purpose was now. He was telling me that I didn't have to know. He knew, and He would fulfill it. *Okay, God. What is my purpose? You'll fulfill it, but what am I going to do? Will I be able to work as a medic again? What am I going to do if I can't?*

Then He told me that He was going to do something that I wouldn't believe, even if He were to tell me. Hope: I felt glimmers of hope. And then doubt would fill me with fear. I broke down Philippians so I could make a conscious decision every time doubt and fear joined together to give birth to anxiety, something I had never really experienced before.

- Don't be anxious about anything. Period.
- In every situation, present or give my requests to God. Hmm, did I have a request? When I thought about this, I realized I was only lamenting and bitterly questioning. I began asking Him to open my heart to hear what He needed me to learn.
- Praying is how we communicate, both talking and listening, with God. This is how we build a comfortable relationship with Him. So in every situation, I need to pray and present my requests to God.

- But first! We go to Him with thanksgiving or by giving thanks. Thanks come from a place of gratitude.
- When we enter God's presence, give thanks, and then present our requests prayerfully, His peace, which is so incredible we cannot even understand, will guard our hearts and our minds.
- Notice that it doesn't say present your requests and they shall be granted. It says His peace will guard our hearts and minds. When He guards these things, He is keeping Satan from wreaking havoc in them. We must cling to this verse so we feel peace.

Proverbs 3:5–8 is one that we read often and was on our fridge, so it was in our faces. But I heard this many times a day in response to my cries. So I broke this down too. I needed things simplified.

- Trust God, not myself, with all my heart. This is both an action and a choice. I was so heartbroken, scared, and hurt that I didn't trust. But as He revealed His provision and quietly spoke His words to me every day, I began to trust. I struggle with guarding my own heart, and when I do, I don't give all my heart to anybody, including God. This is one of the reasons I began begging God to heal me from past hurts so I could stop guarding my heart.
- Lean not on my own understanding. That was good! I had no understanding of my own this season.
- In all my ways, acknowledge Him: standing in His presence to present a request means I couldn't do it with bitterness. If I wanted to know what path I needed to be on, I had to learn to acknowledge him humbly and with gratitude.
- I added verses that aren't as familiar but I needed to hear: if I wanted health for my body and nourishment, then I needed to stop trying to do this by myself since I wasn't wise to the plan and my new purpose.

Another road map and answers to my cries. The four months post-surgery forced me to sit and be quiet because I couldn't do anything else. God knew I would need those months of pause to give me the solitude to cry and then sit still with God. Something I was always too busy to do before.

Of all these verses, Jeremiah is the one I said to myself a hundred times a day and still say many times daily to this day. It is the one set of verses that changed my thinking. When you read it, do you notice that our natural tendency is to put the emphasis on the word *have*? For I know the plans I have for you. But in my despair, as I was crying out, "I don't know what I'm going to do! What am I going to do if I can't go back to work?" I began hearing God say it to me in a gentle, deeply loving way that still moves me to this day:

I know.

I have *plans for you.* I know *your future. You have hope. I will prosper you. You don't have to figure it out.* I *already have.*

This is faith—believing in things unseen. If I let God take my hand, He will lead me to my future. Walk me into those things I wouldn't believe, even if He told me. He will show me. Thank you, Lord, for speaking this truth to me when I so desperately needed it! This leads me to where I am today. At the water, writing my story, healing.

I cannot leave this season without touching on the physical healing of injuries and surgeries. I don't recall growing up accident-prone. Somehow, I blundered into this trait in adulthood. It wasn't because I was reckless or daring; it was mostly because I was thinking about things I needed to do and not paying enough attention to what I was currently doing. Which then altered what I had been thinking about doing due to being injured, so I would've been better off just focusing in the moment. It is a miracle I retired with all my fingers, and the fire service didn't demand a sacrifice of digits. On-the-job injuries:

- dislocation of my thumb from running into one of my partners.
- two fingers with cuts that brought on an infection surrounding one tendon, then trigger finger—stitches to both

50

fingers, surgery on one to clean infection, release of trigger finger. All this from slamming my hand in an ambulance compartment door while in the ER bay. Apparently, my left hand isn't as fast as my right hand.

- puncture from a knife while cutting a watermelon (just trying to be nice for the partners who were out mowing in the heat). The wound sealed itself, and I became infected so badly that it was lanced in the office but the doctor was prepared to take me to surgery. This was a couple days before we drove to NYC. The doctor wasn't comfortable with me traveling, so I had to stop every hour to soak it and redress it. He also wanted me to FaceTime him a couple times so he could look at it. Before I left home, he told me to find a doctor there that would operate on it if it didn't get better, which I did thanks to the help of a doctor on the task force.
- dislocated finger from a training exercise.
- my thumb—Matt's favorite injury. We were dispatched for an unconscious/not breathing overdose. I climbed in the back seat of the fire truck. This was an older fire truck with the doors opening onto the crosswalk. I used my left hand to balance while I reached out to grab the door handle. I swung the door shut, sat back, and felt a tug on my left arm. It was fall, so I was wearing a pullover. I pulled back, assuming my pullover got caught in the hinge. When it didn't pull free, I looked and realized, with horror, that my thumb was caught in the flat hinge! The door was wedged shut, so it didn't easily open. I was kicking the door to get it open in a tight space with very little kickback room, with my left arm stretched all the way out, my right barely reaching the latch, and my steel toe boots kicking the metal plate on the door, making a very loud raucous sound. Now my partners are making it to the truck, asking me what I'm doing. The door flies open, my thumb is released, and I fall back into my seat, looking at my thumb—amazed it was even still attached and disgusted by the sight of it. It was bleed-

ing and so flat that it looked like a cartoon. It was numb, and the guys asked if I wanted to stay behind. I was about to pass out and knew if I climbed out, I wouldn't make it, would fall off the truck, and then have a head injury. I told them to just go on the run. We had a slide window between the back doors, and I had the window open. I was on the floor with my head hanging out like a dog, trying to get enough cool air to not pass out. I heard a running commentary from my partner in the back, telling the guys in the front what he's witnessing. It was like a three-ring circus. One of the chiefs came to the scene to take me to the ER. The feeling came back to my thumb on the way there, and it was bad! I was seriously worried they'd have to amputate my thumb. A crush injury from the knuckle to the tip, a break through the joint, and a total avulsion of the crush area required your help with dressing changes. And those were pretty grotesque. I do believe, Matty, that you might be able to handle gross images now had you not been forced to tend to this injury. You are welcome.

• Four broken noses: wrong place during an EMT class scenario found my nose in the path of a student's finger—that was surgical fix number one. Exactly one year later, Ali was sitting on my lap and headbutted me; I literally saw stars, but my eyes were wide open, and I felt air hitting under my cheekbone when I breathed out my nose. That shattered my nose and was surgical fix number two. The third one I can thank myself for is pulling the overhead trunk door of our Journey down on my nose. That one required stitches too. Over twenty-four hours later, when I got up at the fire department to go home at the end of the shift, both eyes were swollen mostly shut and were black. Hmmm, I guess riding the bike for an hour wasn't any better for me than having run on the treadmill. The fourth was also my doing. We were in our old reserve engine, and I was having trouble getting an appliance out of a compartment on a fire. My T. rex arms didn't help matters, as I had my face close to the edge of the compartment to reach the awkward placement of the appliance. When it finally came free, the sudden weight of it rapidly pulled me forward, slamming my face, and more specifically, my nose, into the compartment. That one caused a ridge to form on the bridge of my nose and another black eye.

- More than a few trips to the ER while on duty for debris in my eyes cannot outshine squirting gritty hand soap in my own eye. Who needs enemies when I have myself? Ali barely had her driving permit and had to drive us to the ER. That took a while to flush out. I was still pulling grit out of my eye for days. You guys were ER rock stars! You had the brilliant idea to use an old diaper bag as a "survival" bag. Ali, ever practical, put actual first aid supplies in there since she was always my first line of defense for any at-home injuries. As I had many of them, Matt added small toys and snacks because he got tired of missing meals. We still laugh about the toy pistol he added because, why not?

- When my back surgeon released me to walk the treadmill, in typical Haggard fashion, I went all out and overdid it, causing plantar fasciitis and a heel spur to develop. And, because ignoring my back issue for so long worked out so well for me, I decided if I ignored the pain, it would just magically go away. So limp, I did…again. Treatments helped a little, but when I tripped up a step because I still wasn't accustomed to my left leg deficits and tore my plantar tendon, I knew surgery was in my future. But I needed to wait until the end of the racing season so I wouldn't lose income. So more limping and pain were endured for months before foot and calf surgery (to lengthen my calf muscle).

- I'm only going to touch on the healing after our accident, as I will dig deeper into that in the next book I am writing. You know how incredibly impatient I am when I am healing. I know it slows my progress. I was in awe of your peaceful endurance of pain and learning to be in a wheelchair, then use a walker, Matt. When I would hurt you during transitional moves, I would cringe and apologize, and you would so gently and lovingly tell me it was okay. I didn't envy the situation you were in, but so often, I would contemplate how I wanted to be more like you. You brought heaven down and were, and still are, a living example of grace and love.

I do not want to live with the past hurts of men affecting my current and future relationships. I cannot change what happened. I cannot change how many times hate has shattered me. I am more than my past. But I cannot fix myself. Only God can. The past few months, I've been on my knees, crying out to God to heal me. I have known for months now that God wants me to write my story. I have ignored it. But in the quiet, God gently answered, "The healing is in the writing." If I truly want to heal, and I do, then I must write. I believe your healing is in this writing too. My sweet babies, as you walk through my story with me, feel the great physician's touch on you. He wants to heal you too!

CHAPTER 11

A Time to Tear Down

Many are the plans in a person's heart, but it
is the Lord's purpose that prevails.
—Proverbs 19:21 NIV

This season brings to mind the power of our tongue. It was my duty to raise you with encouragement, gentle discipline, and guidance. Too often, my tongue used words, when coupled with my tone and body language, to criticize you and tear you down. I deeply wish I could've handled myself better, in a way that did not cause you distress. I never set out to be that way, and I did not like myself in those moments. God's grace worked in us, and forgiveness isn't just words or an action to me; it is your beautiful faces and hearts. I pray that you forgive me for every hurtful word and every harsh thing I said. I believe God used those years—not just my poor behavior but our cord building—as a way to define all three of our purposes. Each one of us is assigned a very specific purpose, and when I stopped struggling to figure out how to do it my own way, I found His purpose will prevail. Thank You, Lord!

CHAPTER 12

A Time to Build

Forget the former things; do not dwell on the past. See, I am doing a new thing! Now it springs up; do you not perceive it? I am making a way in the wilderness and streams in the wasteland.
—Isaiah 43:18–19

Building always happened after a life marker for me: after my dad died, after our accident, and after retiring from the fire department. This is because each marker represents when I fell apart. If a carpenter had had to actually put me back together, hammers, nails, nail guns, saws, glue, clamps, and bracing would've been used. How painful! Surgeons have actually used some of these tools to literally piece me back together. After each shattering event, God held my face so I would look into His gentle eyes and said to me, *Don't look back; fix your eyes on Me; look instead at what I am doing for you—something new! I am making a way for you, a way filled with life and beauty.* And He did. I loved my dad, but his alcoholism was destroying our family. The new thing was not easy, but it was peaceful. The accident and all that transpired to change our lives afterward were not easy, but the new thing strengthened our relationship and our new purpose in life emerged. The two-year transition leading into retirement—Surfside deployment, COVID-19, menopause, back surgery, foot and calf surgery, and selling our house—was not easy. But the new thing led me to living with my best friend for that year and a half, buying the perfect house with my parents and Eric, and planning my wedding.

Just a couple weeks after returning from Surfside, recovering from a moderate case of COVID-19, and feeling very weak and exhausted, I just wanted out of the house, so Eric took me on a motorcycle ride. I did not yet know about the spinal cyst and what huge changes were in store for me. It was the first of August 2021, and it was a beautiful day—sunny, bright blue sky with just a couple clouds. I was praying and distinctly remember asking God to please just build me back to where I was. Immediately, I heard His words tell me, *I am not going to build you* back. *I am going to build you* forward. What! What did that even mean? I know I didn't make that thought up because I wouldn't have said it like that. What does building me *forward* mean? I pondered this statement in the days and weeks to come and still didn't understand what it meant. With the realization that I wouldn't be going back to the fire department and the task force, this statement now became a promise that I clung to.

The body that I had pushed to the max for years, made unbelievable demands of, that carried me into and through fires, collapsed structures, and miles of disaster debris, successfully carried both babies while making these demands, and healed quickly through every ridiculous injury and surgery, was no more. God wasn't going to build me back because I was not going back. This is still something that I am coming to terms with. I miss my old body. And now I am left with a literal old body. I do not need the body that carried me through what I once did. God is still building the body that I need to carry me through what is ahead of me, looking forward. He made a promise to build me before I knew I needed building. What a gracious gift that was...is! Building is painful. It is not easy. But when I stop struggling to take the tools from the carpenter who has the master plans, what takes shape is a beauty to behold.

CHAPTER 13

A Time to Weep

Listen to my words, Lord, consider my lament. Hear my
cry for help, my King and my God, for to you I pray.
—Psalms 5:1–2 NIV

For he has not despised or scorned the suffering of
the afflicted one; he has not hidden his face from
him but has listened to his cry for help.
—Psalms 22:24 NIV

Scorn is not a word we use much. It means to reject or refuse. We know what rejection feels like. You lived through that with your dad, and ultimately, when he decided he did not want you in his life anymore, he rejected you—your love. He refused to have anything more to do with you. Your dad scorned you. But God, your Father, will not scorn you. He hears our cries for help, whether for something small or really big, poured out to Him in a long lament or a deep begging cry for help.

The shortest and deepest, most heartfelt cry for help that I ever offered up to Him happened on July 24, 2005. I was kneeling on the inside roof of our crashed car, and I was having a desperate, out-loud debate with myself on whether I should get you out of your car seat, Matt, or leave you and let the fire department get you out. When I knew with a knowing in every cell of my being that I was about to lose you, I cried out, "No! God! Please! No! Help me!" In the next year to come, I truly began to see how incredibly He answered my cry as you began to reveal what I knew when I cried for help. You were gone. God came down to get you, and you experienced peace beyond our understanding. While you were asking God to not come back, I was imploring Him to help me. Your soul weeps in this season because you just want to go back home to heaven. I cannot envision life here without you. Only by God's miraculous hand and graciousness did I escape an eternal season of weeping.

CHAPTER 14

A Time to Laugh

Laughter has always been the healthy dose of medicine we needed. We laugh at ourselves and tell others what we did or what happened so they can laugh. When I'm really nervous, I laugh uncontrollably. There are a lot of times when it is not the right time to laugh. Laughter is a release of emotion, it can be a gift we give to others, and it can hurt when done to someone. Law enforcement, EMS, and fire use laughter as a means to let go of what we just saw, did, or encountered. It seems sick, but it's actually therapeutic. I've not found scripture on laughter that applies to our lives, but I did find that when we were troubled and turned to other scripture God gave us, laughter from a lifted heart soon followed.

While most of my career in the fire service was troubled, I remember quite a bit of laughter. Just as we chose to laugh a lot, I chose to find good and not just live the bad. One specific incident comes to mind: it was around 1997 at Station 2. Sparky, our Dalmatian, was still alive and was notorious for eating food off our plates if our meal was interrupted, even if it was covered and on the middle of the table or on the counter. We had taco salads for dinner, and leftovers were put away. Sam, with a full belly, kicked back in the recliner and fell asleep—snoring asleep. I thought it would be funny to lay a few tortilla chips in his lap, not even considering Sparky might see and be interested. And interested he was! He started sniffing around Sam's lap and carefully ate a couple chips. Zach and I looked at each other with big eyes, a wicked grin, and barely contained laughter, the same idea forming in our heads. We quietly went to the fridge, got all the leftovers out, and carefully built a taco salad on Sam's lap—meat, beans, salsa, sour cream, and cheese—on top of the tortilla chips that hadn't been eaten already. The snoring never changed. Sparky began eating and, in order to get the best angle, climbed up in the recliner.

Sam woke to Sparky straddling him on the recliner, his nose burrowing on his lap, nudging him, and licking his pants. Sam pushed him off, calling him names. He still had no idea about the lap salad, and we were hidden around the corner, dying laughing! Sam promptly dozed back off after huffing and grumbling about how dumb Sparky was. And as we were coming around the corner, Sam sat up with a confused look on his face, sniffing and asking us what the smell was. We were looking at each other and pretending like we didn't know what he was talking about. He then sat up, still sniffing, looked down at his lap, which Sparky had licked clean, leaving behind sour cream stains and a wet spot. In a disgusted, loud voice, he wanted to know what was wrong with that dog! He stormed to the locker room to change his uniform pants, only to find the only pair available was an older pair that he had grown out of, so he had to wear pants he could barely fasten while the dirty ones were washed. He had no idea what had happened until we told him much later. It was before cell phones, or there would've been video evidence.

Laughter is a choice. It's like skipping. I've observed children skipping over the years, and never, ever does that child have a frown on their face while skipping. It brings a smile to the face and, usually, laughter to them and to others. I choose to smile. I choose to laugh.

CHAPTER 15

A Time to Mourn

The Lord is close to the brokenhearted and
saves those who are crushed in spirit.
—Psalms 34:18 NIV

Luke Matthew Robin was a teeny little bird that fell out of its nest, and I found him when I was mowing. I called you guys out to look, and you wanted to save it. Ali, I don't recall how invested you were in saving a bird, which obviously only had minutes to live, but the name clearly indicates who was fully vested. Luke Robin was put in a box that was as comfortable as it could be and watched over the rest of that day. We prayed for him, and I really don't recall what nursing measures were given, but (I'm sorry, Matty, I don't remember every minute detail) I think he was still alive in the morning, but not for long. Matt took this so hard! I watched you dig a hole in the corner of the fence; you made a little headstone, and then Ali and I were called out for the funeral service. You were so serious. You spoke kind words about what Luke Robin meant to you. Then you asked Ali to say a few words (the sidelong glances between us—I don't even remember if she spoke any). Then it was my time. You cried as you covered Luke. Something about this little bird struck a deep chord in you. I know if you ever have a son, he will be given this name.

We have much to mourn in this broken world. God is always near. Specifically, when our hearts break, no matter how trivial it seems to others. When our spirit is so crushed, we feel like an anchor is tied to our heart, our Lord saves us.

I love kids. They are a safe place for me. When I'm in a situation that is awkward for me, I look for kids that I can hang out with. They accept me; I can be silly and laugh with them. While I hated the need to transport a child, I loved that I had the privilege to care for them. In this season, two different spontaneous abortions came to mind. M22 was called to assist a volunteer department. I was the first to arrive and found a young girl on the bathroom floor with blood all around her and a tiny fetus lying in the middle of the blood. This patient stated she had no idea she was pregnant, had stomach pain, went to the bathroom, and the baby fell out into the toilet. I had only been a medic for less than a year, had no idea what I was walking into, and proceeded to treat the patient as if I had just delivered the baby because they never really covered this in medic class. The baby looked to be around eighteen to twenty weeks old, perfectly formed, and even though no one was expecting nor excited for its arrival, my heart broke for what would never be.

I was no less amazed at how tiny and fully formed the second spontaneous abortion was. M22 was called to assist in 1's area for an OB. This lady knew she was pregnant and did not report any pregnancy-related issues. She had been cramping. She went to the bathroom, and the baby fell out and into the water. She was still on the toilet in extreme distress when I walked in. She felt like the baby was moving, but it was her uterus contracting, and since the baby was still attached by the umbilical cord, every contraction moved the baby. We helped her up and to the cot. I clamped and cut the cord, and I wrapped the baby in a towel. Around twenty to twenty-two weeks, this baby was also not viable. I spent more time looking at this little boy after we dropped the mom off in the ER. Skin translucent, I could see the venous and arterial systems, depicted as blue and red lines; eyelids formed with a slit to separate the two; ten fingers and ten toes; clearly a boy; hair beginning to grow on the body; a round head; and a little ribcage. As I looked at this beautiful boy, my heart ached to think that anyone would knowingly abort a tiny human because they were not planned or wanted and were thrown away like trash.

CHAPTER 16

A Time to Dance

A person's steps are directed by the Lord. How
then can anyone understand their own way?
—Proverbs 20:24 NIV

Do you remember dancing to the *Tarzan* soundtrack? Similar to
skipping, dancing makes us smile. Those are some of my fondest
memories. We danced freely, singing loudly. Ali, do you remember
the video we sent you your freshman year of college? You were home-
sick, and we missed you, so Matt and I decided to send you a video of
us dancing and singing to a song I can't remember. It made us laugh,
and we knew it would make you laugh too. I know Matt still has
the video. It shows the lengths we go to make sure the strand that is
struggling is strengthened.

Dancing with Ms. Beth was an actual thing when I subbed in
elementary school. Every morning, we watched video songs, and most
of the kids would dance. There was one particular girl who really
loved to dance, and she grabbed my hand to join her one morning. I
made the other kids laugh, and they then wanted to dance with me
too. Each morning, if they had been good so far, they were rewarded
with dancing with Ms. Beth.

It's interesting that dancing is paired with mourning. Maybe
movement loosens us up so we are able to accept God's presence.

Maybe when we're dancing, we put down the stones we're using to build walls to hide behind. I think moving our muscles loosens us up, so we don't just sit and tighten into a ball.

CHAPTER 17

A Time to Scatter Stones

Light in a messenger's eyes brings joy to the heart,
and good news give health to the bones.
—Proverbs 15:30 NIV

When I think of scattering stones, I think of what stones scattered on water do. It creates ripples. I refuse to believe the many men throughout my years in the fire service who felt the need to tell me to "shut up! No one cares! If anyone cared what you had to say, we would ask!" were correct. They did not know me. They ignored the people I impacted—those I made a difference for. The number of patients and family members who either showed up for my retirement party or sent a message to me is a testament that what I did mattered. Generations carried on; people got more time with their loved ones—all the people my former EMT students impacted.

People like to call firefighters heroes. We are quick to shake our heads no and disagree: "We're just doing our jobs." I had a profound realization the night the 9/11 Memorial and Museum opened. I was in NYC for a task force class and was housed in the hotel across the street from the 9/11 Memorial and Museum. On May 15, 2014, the museum was dedicated and opened twenty-four hours a day for a week for all survivors, family members, and those who worked at ground zero. One of my teammates struck up a conversation with one of the architects, who said we could go. The three of us went in our uniforms around 10:00 p.m., and it was crowded. And yet, as quiet as the Holocaust Museum, reverent. We all went our separate ways. I spent time in the gallery that had pictures of FDNY working the pile before USAR arrived to help. Even though I'd been on the task force for nine years at this point, deployed many times, and had been to many training exercises, it really hit me that at any time that could be me standing on a pile like that.

Ironically, seven years later, that *would* be me in Surfside—my last deployment ever. We had a family vacation to NYC scheduled for a couple weeks later, with plans and tickets to visit the museum. I knew this room was where you would start fully understanding what I could encounter. Even though you had been to many training exercises and been utilized as patients many times, I knew you didn't understand what I may be called to do. There is a room with pictures of every victim, where you can look up their obituary, and it is read while pictures show on a screen. As I walked through there, I heard weeping and recognized that those were family members. I quietly left that room. As I was looking at a large picture of a firefighter, a young lady and her two young daughters came up behind me, and the lady took a sharp breath in and then told her daughters, "That's your daddy!" I respectfully moved away. I began to feel like I did not belong there.

The Shanksville room left me sobbing. In this room, audio clips of people on the plane that were left on voicemails and answering machines were played while a board showed a timeline of that flight. There are benches in there. On the front bench was a young man who leaned forward with his elbows on his knees and an older woman right next to him with her arm around his shoulder and her head leaning on his. The man was sobbing—shoulder-shaking sobs. I was overcome with pain for them. Obviously, they weren't survivors and had lost someone on that flight. As I went upstairs to leave, I was stopped by a man and his group. He said, "Indiana! I remember you! I can't tell you what it meant to us to see people come all the way from Indiana to help us. Thank you. Thank you." They shook my hand. They were FDNY firefighters. While I felt like an impostor because I wasn't there, I represented my team, which was. At that time, the memorials weren't blocked off after hours. It was almost 1:00 a.m. I stood by myself at one of the memorials, envisioning the enormous debris pile that had been at this site, and realized, sadly, for the first time since I began in the fire service, that what we do *is* extraordinary. We think it's just what we do—just ordinary. But when you break the word *extraordinary* down, it is extra ordinary. It's more. I felt bad for every time I downplayed what we do as nothing when someone wanted to thank us. They were saying we are more than ordinary, and they were correct. I committed at that moment to simply say thank you to anyone who thanked me for my service.

I made a conscious choice to do my job. Yes, I made sacrifices. But you both sacrificed because it was exacted from you. You didn't sign up to sacrifice; you were never given a choice. Every time I was deployed, I was doing what I loved to do, and you were left living through the disappointment of a missed trip to NYC, cancelled the night before we left; Ali's birthday gift of tickets to Pentatonix with me changed to taking Matt and sending me a video; each of you missed out on birthday plans, holidays, school events, and doctors' appointments because I was gone. The fire department was similar, but at least I tried to make it, even if I was pulled away every single time before you even got started. Ali's prom hair and makeup were done at 1's, with pictures in front of the fire trucks. I could only do what I did because I had your support and understanding. I was able to make a difference for many because of the sacrifices you made. I am forever grateful for the support and understanding you both gave me!

CHAPTER 18

A Time to Gather Stones

As a child, I loved to collect rocks. I would just as soon be walking up and down a gravel drive, searching for any pretty, unusual, or interesting rock, as visiting people. I still have rocks that I collected as a child. Just the other day, I found an unusual rock in our driveway and put it on the porch for others to see. I especially loved geodes. The outside of this rock is so plain that it is almost ugly. Unless you know what you're looking at, you wouldn't know that the work it takes to break it open reveals a unique and surprising beauty.

Through all the hate and meanness heaped on me hour after hour, shift after shift, year after year at the fire department. When they were done dishing out hate, I chose to be invisible. It was how I kept from inciting more hated. Being invisible killed me. I often likened myself to a geode. Most people in my life had no idea, no clue, who I was on the inside; how beautiful I could be; or that what I had to offer could add value. I learned to build walls to protect myself. It's not healthy, but I didn't know what else to do. I literally envisioned myself gathering stones and building walls. Building was an action I could do to protect myself.

As the walls went up, haters disappeared behind the strong wall I built and could not access me. I felt safe there, but I did not like being there. I realized years ago that when I walled myself in, I was pushing God away too. Satan took hold of my mind, and I realized I had walled Satan in with me. It was dark and lonely. He took advantage of the hatred of men to encourage me to gather stones and build my walls. He liked alone time with me. It's where he did some of his best work. I am working very hard at speaking up when something hurts me, defending myself and letting no man devalue me because I'm a woman, and speaking for what I need and will not tolerate. This is part of my healing—voicing the trillion thoughts in my head when I go silent from hurt. Stones are best left for me to find in the driveway.

CHAPTER 19

A Time to Embrace

I'm a hugger. I believe hugs heal. It is a gift that is freely and easily given that costs me nothing but has deep rewards for those I hug. It speaks where words fall short. It's the same with a smile. *Embrace* means to hold and support. While a hug can be quick, limp-armed, and have minimal body contact, an embrace wraps the arms around and holds. It can also mean to theoretically embrace something—change, the future, a challenge, "the suck." Gathering crops, especially before automated equipment was invented, would be "embracing the suck."

In medic school, we did rotations with the chaplains, and we talked about the cycles involved in the grieving process, but there was not much to prepare us for comforting devastated family members. As I've said before, as the probie, it fell on me to be the bearer of bad news. No one told me this. I figured it out: when we were finished with our tasks, I would turn around and find myself alone. My partners would be standing at the truck or sitting in it. In the beginning, I had no idea what I was supposed to do or say. I found that when I was quietly and gently frank, delivering the facts in simple terms with kind and understanding eyes, it calmed the family down. Then I realized they had no idea what the next steps were. So I found these out and began guiding them in the very next steps and calling someone to come be with them. Then I stayed with them until someone else could come. I sat with them, I listened, and I embraced. I know how powerful touch is—being acknowledged—and how an embrace tells the one encompassed that they are not alone.

I raised you to be comfortable accepting embraces from those you trust and love. We hugged, embraced, and snuggled as often as possible. To this day, a healing activity for us is to snuggle in bed and watch episodes of *Friends* together. It did not matter how terrible life was; when we snuggled, we knew everything was going to be okay.

CHAPTER 20

A Time to Refrain from Embrace

I tend to hold on to things. I don't horde physical items but things that are familiar, even when I'm all too aware that those things are hurting me. I do not embrace change at all. I'm the exact opposite, actually. I embrace what I know, holding on tightly and guarding it with my body, all while beating away change. My heart becomes troubled when I embrace what is no longer meant for me, when instead, if my eyes are on God, I will feel the peace of letting go. If I allow God to hold my right hand, I can't strongly embrace what He is trying to move me from.

Nothing embodies the absence of hugging more than the COVID-19 era. What an unbelievable, surreal time that was. Like those who were alive for Pearl Harbor and the September 11 attacks, they all remember where they were and what they were doing when it happened; everyone has their own COVID-19 story. For our records, so we don't forget this as time passes, I have written down a timeline that I'm going to share here:

- The novel coronavirus (COVID-19) was first identified in the city of Wuhan in China's Hubei Province.
- It spread around the world, with the Indiana State Department of Health confirming the first case in a Hoosier with recent travel on March 6, 2020.

- News of COVID-19 around the world increased while I was visiting my cousin in Frisco, Texas, on February 17–21, 2020.
- We traveled to NYC for vacation (and also to Washington DC and Virginia) on March 8–14, 2020. News reports covering more cases in the US; talk of canceling flights; seeing reports back home of grocery stores being emptied and people weren't able to get food and necessities (toilet paper). I was making plans on how to get home if flights were canceled.
- We decided to go on frequent walks and get as much fresh air as possible.
- A list of what was cancelled:

  FDIC; Summer Olympics (moved to 2021); Mini-marathon and all month of May activities; Indy 500 (moved to September); Kentucky Derby (moved to Labor Day); horse race season (no spectators, only essential people for races); high school and college graduations (you both finished your year for college from home); NBA; MLB postponed; spring breaks for school; and my task force training trip to Israel.

As I believe that human touch expresses what words cannot, avoiding touch and not allowing people to be around those who are important was a terrible tragedy. I'm not going to say much more about COVID-19. It was an obvious example for this season.

CHAPTER 21

A Time to Search

I sought the Lord, and he answered me; he
delivered me from all my fears.

—Psalms 34:4 NIV

He sought God during the days of Zechariah, who
instructed him in the fear of God. As long as he
sought the Lord, God gave him success.

—2 Chronicles 26:5 NIV

As a parent of adult children, I am honored that you come to me and seek my counsel. I can tell you what I believe you need to do. I can put applications or objects in front of you and tell you it's what you should do, but if you are not seeking my opinions or thoughts, you're probably not going to pay any heed to me. It's the same with God. He wants us to choose to seek His counsel and believe that we need His help. We seek and He will answer, deliver, and give us success. Just as you come to me, I pray you always seek Him first.

Most of the time, when people call 911, they know where they are, so we don't need to search for them. Except that one time, M22 was dispatched to assist a neighboring department with a crash in the middle of the night. It was a warm evening, and we pulled up to an inverted Jeep in a bean field with a lot of law enforcement vehicles around as this was a pursuit. There were two patients. My patient was holding onto the roll bar when it rolled, ripping his hand off above the wrist. He was also thrown from the vehicle. No one had seen his hand by the time I was ready to transport the patient, so I left everyone with instructions to search for his hand, get it in a bag, put the bag on ice, and rush it to the hospital in hopes they could reattach it. They looked all over the bean field and the ditch with the help of K-9s. They finally found it hours later, after the day broke, so far away in the ditch; we couldn't imagine it was that far.

Being part of an urban search and rescue team, searching was our purpose. We walked search grids, we searched under rubble, and there are two types of search teams in our search division. There were only two times in my sixteen and a half years of being on most every deployment we had that we searched for a known victim. One is the baby boy at Hurricane Florence in North Carolina. The most memorable was the Surfside building collapse in Miami, Florida.

At the end of June 2021, I was working the horse track. I specifically recall standing outside the chase truck, leaning on a rail, and talking with Eric when I received a phone call notification of activation. "For what?" I asked. There were no hurricanes stirring, and as I don't watch TV, I had no idea there had been a building collapse the previous week. Eric was also activated and was working the track that day. We were to report to HQ in two hours. We left immediately, went to our homes to pack, and went to HQ. We were on buses in less than three hours and drove straight there. We were housed on a Royal Caribbean cruise ship docked to remodel the ship. Placed on day shift, we got a couple of hours sleep before we began the routine that would follow for the next couple of weeks. As a medical specialist, our day starts early so we can do medical checks on everyone before we get on the bus. We got breakfast before walking over a mile to the buses. Drop-off at the site wasn't close, so we had to walk to our briefing site before getting a report from the night crew. To avoid one shift from doing all the work in the sun, shifts were noon to midnight, which was our shift, and midnight to noon. The way it worked out, I would get about two hours of sleep each night.

It was the toughest, most demanding work I have ever experienced. For the first time in my life, I questioned whether I could actually do it. The beating sun, high humidity, long pants and long sleeve shirts, a half-face mask, a mountain of unstable debris to climb, chunks of concrete to pick up and pass along a line of people, five gallon buckets filled with concrete to pass along, lack of sleep, a hurricane that was hitting the Western side of the state that kept circling around and dumping on us so we'd get cooled from the rain, then shiver from being soaked in the wind, then back to work in higher humidity. It was the ultimate test for me. At the same time, I was at the height of spinal cyst growth, in terrible pain, and having to focus on my left leg not giving out. We knew there were victims that needed to be recovered, and we were there to work until every last victim was accounted for.

I've seen every stage of death countless times, but none of us were prepared for what we saw as we slowly uncovered people. We were digging right next to machinery, feeling the debris vibrate. We went through a set of gloves every couple work cycles, if not before, because we were digging with our hands. We were searching through papers and pictures to see if we were near the apartment of an unaccounted-for victim. We helped bring closure to families, who could then lay their loved ones to rest. The toughest deployment I had ever been on proved to be my last. I'm thankful I was able to be part of something that made USAR history.

Briefing at headquarters prior to leaving; Strachan is on my right and Chuck Jones is on my left. (June 20, 2021)

Strachan, Chuck, and I resting during a rainstorm (July 5, 2021)

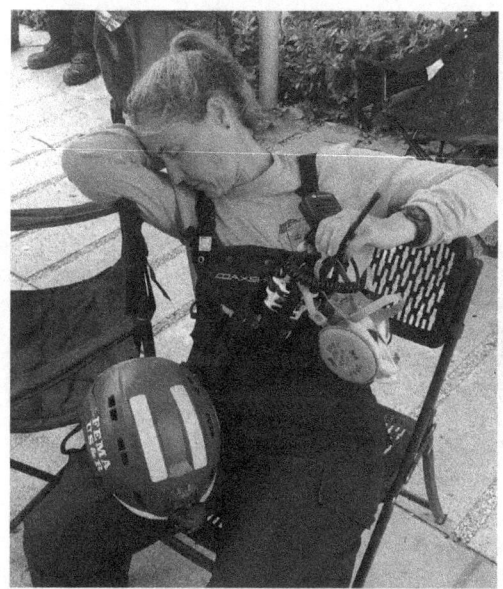

Sleeping during work cycles (July 6, 2021)

Day shift, walking toward an onsite briefing before
beginning our work period (July 6, 2021)

Strachan, Chuck, and I resting in between work periods (July 7, 2021)

National news! Of course, Matt saw this and sent
the picture to me. More rest! (July 10, 2021)

Evening before heading home. Chuck on my right,
Strachan on my left. (July 11, 2021)

Our team, working on the pile. Flying in the air
was debris, and this was very common.

CHAPTER 22

A Time to Give Up

I have been given many scriptures over the years, instructing and encouraging me to not give up. Not once in my prayers did God send scripture my way, telling me to give up. So I don't have any scripture to add here. Closing a door on a season, walking away, turning the other cheek, leaving revenge for the Lord, and retiring isn't giving up. To me, they all mean "it is finished." When I think about giving up, I think about quitting, and I am not a quitter. I didn't raise you to quit. So this season is a tough one for me to reconcile. As I contemplate this, I think of times God would ask me to give up, and it all involves when my refusal to give up becomes not only a hindrance but unhealthy and hurtful for me. As in, it's time to answer my pleading prayers to give me enough strength to get through my remaining years at the fire department, and I, in the very next breath, refuse to leave. If I know God's blessings and plans are always best for me, then why on earth do I dig my heels in and not let go? God knows. He created me to be tenacious, strong-willed, and not quit. Which is why He removes seasons. I have no choice but to accept the new season.

I look to nature and understand how giving up is a wondrous thing. Trees give up their leaves each year and are stripped bare to fill back in and produce again and again. Stems give up their flower petals year after year to be filled with more beauty after a season of winter. A butterfly gives up its safe cocoon to emerge and fly free.

I spent the better part of my early adult years overweight and not feeling good about it. I was at my heaviest when I got pregnant with you, Matty. The most successful weight loss happened not by choice, but due to the constant pain I was in. It killed my appetite. I committed to getting in shape and continuing with the weight loss. Over the years, through proper nutrition, working out, and AdvoCare products, I was in the best shape of my life—strong and fit. I felt good and liked what I saw in the mirror. I was fascinated with the body my hard work created. Maybe I was too fascinated. All it took was hard work and determination. I didn't need God to help with that. I had those traits mastered. Until I didn't.

In early 2021, my activity level slowly decreased due to my left foot dropping because of increasing numbness in my lower leg. I was tripping and stumbling a lot. I was becoming increasingly tired because the pain was keeping me awake at night. I would try to nap instead of working out. Since I had moved at this same time, I didn't have the yard work and garden to tend to, so that activity went away. By the time I was deployed to Surfside at the end of June, I was sitting a lot because, in that position, I could pull my left leg toward my body and gain a little relief. When I came back from the deployment with COVID-19, all I could do was sleep. A few days in, I decided I needed to walk. I set out on a slow stroll of less than a quarter of a mile, and I barely made it. I was hurting, having trouble breathing, and so weak I thought I would collapse. I don't know why I thought that was a good idea when I would be so exhausted and my leg would hurt so bad that I couldn't go fifteen feet without lying on the ground for relief and to gain strength to finish the trip to the bathroom.

One of the weird and unexplained side effects of COVID-19 is how it messes with women's ovaries.

When it took hold of my body, it picked me up and slammed me into menopause—hot flashes, ha! As a firefighter, a flash is intense but here and gone. These weren't flashes! To me, they were hellfire burning, and if air wasn't flowing across every available surface of skin, the gates of hell would open, and hellfire would consume me on a cellular level. My joints were hurting terribly. I desperately longed for and needed sleep. My nights now consisted of setting up a fan on high, lying on my back with my left leg bent as far back at the hip as I could get, trying not to move from that position of relief when my joints started aching so bad I couldn't take it, then tossing and turning to find complete joint relief, which then cranked open the gates of hell to release its fire. I had no idea what was happening, as nothing had been diagnosed yet. Was it something I absorbed from Surfside? (I had enrolled in a University of Miami study for all who worked the Surfside pile since there were extremely high levels of toxins in the air.) Was it COVID-19's aftereffects? I began to gain weight and lose muscle. And I hated it.

Back surgery left me with no core strength and no way to work on that other than holding myself upright, which was exhausting. Walking, a form of therapy for me, became putting one foot in front of the other and standing upright. Still with menopause as yet diagnosed, I began holding weight in my belly. My clothes didn't fit comfortably. So when I was released to walk the treadmill, I went nuts until my foot decided I was meant to sit. No matter how healthy I ate and how small my portions, being sedentary was not my friend. I felt like I was in somebody else's body. I kept saying, "This is not my body!" I began to suspect I was in menopause, and my doctor confirmed it. I began taking medicine that would help and finally got relief from the terrible joint pain. Hot flashes diminished, and I was finally able to get a full night's sleep for the first time in almost two years.

I have had to learn about my body. I am not where I once was. The body that carried me through more than I could've expected, responding every single time to ridiculous demands over the years, did not need to perform at that level anymore. I have to choose every single day, many times on most days, to be thankful for the body that was, the body that has healed completely through an avalanche of trials, and the body that can still carry me on walks and work it.

The season of giving up is a season I am still in. It is the season I have struggled the most with. It is the first season I have endured that I cannot alter or change with my sheer will and determination. It is not ironic that God was telling me to write my story for healing while in this season. God is good, and I thank Him for a healthy, whole body.

CHAPTER 23

A Time to Keep

But you, keep your head in all situations, endure hardship, do the
work of an evangelist, discharge all the duties of your ministry.
—2 Timothy 4:5 NIV

There are a number of verses that instruct us to keep his command-
ments in our hearts; keep the faith. This verse says to keep your head
in all situations. I believe for some people this is a spiritual gift, and
God places them where they can use it for His good. It is also a trait
that can be built and developed.

Since I must analyze everything from every angle so it makes
sense, every hate-filled incident from every single man I had to work
with was deeply analyzed. Did I do something wrong? Did I somehow bring that on myself? Should I have not spoken up? I came to
the conclusion that it was not in any fashion because I was lacking
in my skills and abilities or because I didn't add value to the team
or those I served. So it explained how we could leave a training or
a meeting where I had been unleased upon with hatred and disgust
for breathing air in their space, and they would immediately turn to
me to provide a solution when one was not easily seen. My insides
might've been coiled tight and unsure, but because keeping my head
in all situations was my tool to calm and de-escalate, I was most
often the solution. It's an absolute fact that keeping my head saved
lives because lack of faith in oneself leads to panic, which leads to

quitting. And I don't quit. This instruction to keep my head is the one thread of familiarity I clung to during our accident. It is what allowed me to save your life, Matty. I am deeply moved, humbled, grateful, and thankful that God created this gift, this spirit, in me. Because of this and His answer to my pleading prayer while trapped in the car, you are alive and blessing others with your patient love and kindness.

CHAPTER 24

A Time to Throw Away

So do not throw away your confidence; it will be richly rewarded.
—Hebrews 10:35 NIV

Matty, I think Ali and I could agree that one of the first things to come to mind with the term *throw away* is your lack of desire to do this as a child. I'm laughing as I write this! Can you think of what story I'm recalling right now?

I will start by saying that, as an adult, you have learned to hold the treasures that are dear and meaningful and to live minimally otherwise. But as a child, everything was a treasure: fast-food kid's meal toys, including bones; scraps of paper; and broken toy cars; items I threw away as trash were recovered and deemed treasure worth keeping. I decided one day you were going to assist me in cleaning out the drawers in your room. I dragged the trash can next to your dressers and began methodically going through every single item and either throwing away all I believed not worthy of keeping or giving them to you to do. I should've realized the trash can wasn't filling up and needed to be changed. It wasn't until we were almost done that I realized you were either placing what I was handing to you back in an already cleaned-out drawer or digging it out of the trash. Treasures are not to be thrown away. I learned that if I believed something needed to be thrown out, I should do it while the trash can was at the curb and the trash truck was in sight.

While this season is about knowing when to throw something away, I added this verse to remind us all that our confidence is not to be thrown away. Satan doesn't want us to be confident. Confidence gives our mind the clarity to do what's right, even when no one is looking; confidence calms our spirits when everything is falling apart around us; confidence steadies our hands that shake from nerves when faced with daunting tasks; confidence gives us the strength to hold our heads high when the weight of hate seems too much; confidence gives us the courage to look people in the eye and truly see them; confidence keeps our knees from buckling when overwhelming fear presses us down; confidence roots our feet when everything in us says to walk or run away; confidence gives us our firm foundation while things and people in our lives are crumbling away. Arrogance is Satan's replacement for confidence that is thrown away. Confidence is richly rewarded by God,. It doesn't need words for proof. Confidence clothes us with calm and clarity.

CHAPTER 25

A Time to Tear

One definition of a *tear* is a hole or split in something caused by it having been pulled apart forcefully. When your dad decided that he didn't want either of you in his life anymore, understandably, you clung to the memory of what was. The longer you worked to maintain some sort of relationship, the harder he worked at showing you he was done. This tore a hole in the fabric of your being because you loved him unconditionally, and he rejected it.

Sometimes, as we'll see in the next season, there are times to mend. Ali, you and I had hours of conversation spanning months about whether to take your dad to court to hold him responsible for our divorce decree and his promise to help pay for your college. Our cord prayed about it, and all three of us believed that God was directing us to do it. We hired an attorney but could not come to an agreement, so a court date was set. We continued to pray and continued to believe we were doing as we were directed. As your dad and I were both in the same fire department, working at the same station on different shifts, his unnecessary explanation of what was happening to everyone not on my shift was not at all truthful. Hate and animosity for me grew. I was not surprised by the lies and deception your dad was swearing on, but it was eye-opening and unbelievable to you. This was a very low point for us, and I realized the week before we went to court that we needed help or there was no way we would be capable of representing ourselves the way we wanted and needed to. As this battle was taking a lot of funding, I couldn't afford a counselor, and how do I go about finding the right one anyway? God sent us to the best EAP counselor for us. We began seeing him together and were able to meet with him a couple times before we went to court, and then for weeks afterward.

There's not much I care to say about our day in court that will not come off as whining and complaining. I will say about that unbelievable day:

- You spoke on behalf of yourself and in defense of yourself calmly, honestly, and maturely.
- You listened to every word your dad said with no preconceived prejudices, just fully hearing, for the first time, what he truly felt and the lies he uttered to build his case.
- You didn't regret fighting for what you strongly believed in and for the truth to be told.
- Even though it ended up costing me more to fight than to just fully pay for your college myself, we did what God directed us to do.
- We believe that had we not endured that day and the subsequent ruling against both of us, you wouldn't have believed that your dad wanted to cut you out of his life so completely. You needed this to move on and never give your dad future access to pulling you back into his lies and further hurting you.
- Money is your dad's god. He walked out of court, believing he had won, but he didn't. He lost. Our God strengthened our cord through that lost court battle; He strengthened us in a way that not a single thing on earth could undo.

The tearing of your relationship with your dad was and continues to be a good and healthy tear. God provides, so I can provide. I am so proud of you both for showing your dad what love is.

CHAPTER 26

A Time to Mend

To *mend* something is to repair something broken or damaged. I think sometimes mend and heal are used synonymously. But mending must come before true healing. Will a cut heal without being mended? Sure, it can. It's probably not going to heal as quickly and painlessly as if it were mended.

I appreciate tape. It was a favorite tool of choice in my years as a medic. As a child, I used tape once to piece together a picture frame that had broken when my brother and I slammed into the wall it was hanging on during a fight. It worked in the moment, but it wasn't the proper tool to repair or mend the damage. Had we used glue or a nail and hammer, Mom might not have known what we did and why. Of course, that would require us to not fight long enough to accomplish that goal, and, well, no.

When you visit a new doctor's office, you have to fill out your medical history and list surgeries. I've been broken so many times that it's hard for me to remember all the brokenness, much less when it was mended. That's the part of the new patient packet that takes me the longest. I've been repaired nicely. Except that one time, when I leaked cerebrospinal fluid internally from a spine injection during my hysterectomy and the blood patch attempt by the anesthesiologist left me leaking like a pin cushion, causing permanent nerve damage from trying unsuccessfully to get an epidural in. My back spasms a lot if I'm not constantly moving my back muscles. What one doctor used to mend a specific injury was not the same tool choice for a different doctor for a different injury.

Sometimes we are called to mend relationships. As every relationship and individual is different, each mending is different. If I was wrong and messed up the relationship, I am certainly more than capable of continuing to mess it up. When we go to God, our Father, and humbly ask for direction, the right heart, the right words, and the right time, He will guide us in a way that is not only best for us but also for the other person.

I wrote the following the first time I went to stay with you after my back surgery. I was withdrawn, depressed, and still crying every day. I had brought the notebook I write scriptures in. I poured my desperate thoughts onto the paper: "Lord, I'm scared. Heartbroken. I feel alone. I don't know where I'm going. Be near me; please don't leave me." I began reading the verses, feeling a calm that stilled my troubled soul. On a blank sheet, I quickly wrote verse after verse of promises. As I looked at what I had written, phrases and words jumped out at me: do not be anxious; be still; trust the Lord with all your heart; I know the plans I have for you; the Lord is close; the Lord goes before me; the Lord holds my right hand; do not fear; I am doing something new; I am doing something you wouldn't believe.

And then I saw myself walking with the Lord beside me, holding my hand, and He was saying to me, *I will keep hold of your right hand. I will go with you. I will not leave you. I am close to you, so be still. Do not fear. Do not be anxious about anything. Trust Me with all your heart. See, I know the plans I have for you—something you wouldn't believe, even if someone told you about it. I am doing something new. You do not have to understand. Acknowledge Me in all you do, and I will make your path straight. You will feel My peace, and it will guard your heart. I am making a way. Do not dwell on the past. It is time to give that up. Time to mend. Time to heal. Time to laugh. Time for peace. I am not going to leave you. I will give you strength.*

I was humbled and deeply grateful that He gave me this vision. My mind was not all rainbows and sunshine after that. I continued to wrestle with Satan in my mind for months and months after this vision. My weapon of choice to fight him was, and continues to be to this day, God's word. This was the turning point for me. The vision and words that mended my shattered heart so I could begin to heal.

CHAPTER 27

A Time to Be Silent

Be still before the Lord and wait patiently for him;
do not fret when people succeed in their ways,
when they carry out their wicked schemes.
—Psalms 37:7 NIV

Wait for the Lord; be strong and take heart and wait for the Lord.
—Psalms 27:14 NIV

He says, "Be still, and know that I am God; I will be exalted
among the nations, I will be exalted in the earth."
—Psalms 46:10 NIV

The Lord is in his holy temple; let all the earth be silent before him.
—Habakkuk 2:20 NIV

I added verses on being still and waiting. Because my inherent nature is to be in nonstop movement, when I am asked to be completely still, I am then silent. God commands us to be still and wait patiently for him. He also commands us to do this before the Lord. We must come into His presence in order to be before Him. We must make a conscious decision to go before our maker and be still—our body, mind, soul and spirit...still. And then we wait patiently. I do not believe, nor do I envision my mouth running, because I am in a posture of humble awe and gratitude. Be still, wait patiently, do not fret, be strong, take heart, know He is God and you are not, and be silent. As I type these and repeat them silently, my head is bowed, and I feel peace.

Silence became a defense mechanism for me through the years. If I kept my thoughts to myself and my face was impassive, those whose sport it was to make me suffer couldn't know how they were affecting me, and maybe they wouldn't do it again, believing it didn't work. So silence became my weapon of defense. I don't advocate this, and I know that it is not a healthy response. I just didn't know what to do to protect myself. Silence can destroy relationships. Since I am no longer required to be around hate, any hurtful actions or words are inadvertently done by those who care about me. I immediately withdraw inside myself and become silent. I cannot get my lips to unseal in order to speak in my defense. I have prayed deeply about this. It is a big part of my cries to God to heal me from past wrongs so it does not negatively affect current and future relationships. I thought God would soften my lips to open and let my words spill forth. But instead, He kept telling me, "The healing is in the writing." Maybe this book is so those important to me can understand my story, where I am emotionally, and where I hope and pray I can move to. Also, the healing is for me and both of you.

God called all the earth to be silent. In a chaotic, loud world, I cannot imagine complete and total silence. Sitting outside our house in the country at night, chaos is silenced, but there is still a symphony of locusts and crickets, coyotes howling, dogs barking in the distance, and corn rustling in the breeze. If God can command the earth itself to be silent, I reckon I should be too.

CHAPTER 28

A Time to Speak

As a brand-new paramedic and the youngest member of our fire department, I felt like I needed to have it all together and not let anyone, including patients, see me nervous. But inside? When the tones would drop, in an instant, my heart rate would increase, my breathing would become so shallow that I felt like my lungs had shrunk to the size of walnuts, my nostrils would flare to get every molecule of oxygen in my walnut lungs, my jaw would clench, and my bladder would spasm, indicating the desperate need to empty, even if I was walking out of the bathroom when the tones dropped. But my face showed complete calm. Panicked, crying patients would increase my nerves, so I very quickly learned that if I could de-escalate the patient, family, and/or bystanders, my nerves would lessen and I could think clearly. I learned that by speaking my belief in what was going on and mapping out the treatment plan, I calmed and gave patients and their families confidence in me, which gave me confidence in myself. Once I had begun the treatment plan, I began to say, "Tell me your story!" and then I would start asking questions. I realized that brains focused on answering questions tended not to focus on what was going on; talking about their lives and what mattered to them—their careers, travels, pets, family, extracurricular activities, favorite pastimes—me expressing true interest and interacting with them calmed us both. By talking with the patient, I learned so many fascinating stories and learned to appreciate the sacrifices people had made for family and country.

Notice I just said that by talking with the patient. To speak is to use my voice to say words. To talk is to converse. There is a time for both. At one point early in my career, we had a fire chief who was terrible. He used his position to harass and the fire department as his playground, and he was the playground bully. He had not one goal to do good for those who trusted us. I spoke out freely in defense of a couple of fellow firefighters who stood their ground for doing what was right. For the four years he was chief, I was punished for that: laughed at, ridiculed, called terrible names, screamed at, followed around town, received harassing phone calls, and called into his office no less than twice each shift, where he would slam the door, lock the door, close the blinds, and stand in front of the door so I couldn't escape, all while screaming at me and threatening me. It escalated in the bay one day when he was screaming in my face, calling me names, and furious that I was defending myself. He drew his fist back and began to bring it down toward my face when he caught himself and stopped. This behavior was passed down to many weak-minded bullies in the ranks for years. To the point that men newly hired in the fire department would hate me and have that same demeaning attitude for no other reason other than the culture that was passed on to them. I continued to speak for what was right and wrong and tried to get anyone who could do something about him to hear me and believe something so outrageous was actually happening. I was pregnant with you, Matty, when I filed a lawsuit and won against him.

CHAPTER 29

A Time to Love

Though one may be overpowered, two can defend
themselves. A cord of three strands is not quickly broken.
—Ecclesiastes 4:12 NIV

What a person desires is unfailing love; better to be poor than a liar.
—Proverbs 19:22 NIV

A new command I give you: Love one another. As I
have loved you, so you must love one another.
—John 13:34 NIV

Be completely humble and gentle; be patient,
bearing with one another in love.
—Ephesians 4:2 NIV

Love must be sincere. Hate what is evil; cling to what is good. Be
devoted to one another in love. Honor one another above yourselves.
—Romans 12:9–10 NIV

Love is patient, love is kind. It does not envy, it does not boast,
it is not proud. It does not dishonor others, it is not self-seeking,
it is not easily angered, it keeps no record of wrongs. Love does

not delight in evil but rejoices with the truth. It always protects, always trusts, always hopes, always perseveres. Love never fails…

—1 Corinthians 13:4–8

I know I've quoted our cord verse before, but love is what binds our cord. Our three strands are not quickly broken because God, who is love, binds us. My greatest gift in life is the love we share. There are a multitude of verses about love in the Bible. When I look at the verses in 1 Corinthians, they fully represent the love you both share with each other. It is a gift freely given to the other for all to see, which then becomes a gift for others.

We use the word love often, but not in the context of these verses. I love my chocolate chip cookies; I love being at the water; I love kitties; I love babies; I love what I do for a living. I have to be careful not to love them so much that they take my focus off of living out true love for others. It is tough to love others when they push our intolerance buttons. God knew this, which is why He gave us guidance: Love is; it is not; it does not; it always; it never fails. I have written this out and laminated it, keeping it handy in my room, to always remind me:

> Love is patient and kind.
> Love is not proud, self-seeking, or easily angered.
> Love does not envy, boast, dishonor others or
>   delight in evil.
> Love keeps no records of wrongs and rejoices
>   with the truth.
> Love always protects, trusts, hopes, and perseveres.
> Love never fails.

I've had the opportunity to pour my love out on my patients, babies, kids, and students in the schools I would substitute in— those scared and purposely hurt by others—as a paramedic for my grandma, my mom, my stepdad, and you.

Not long after you were born, Ali, I was working overtime as the only medic running the entire county. I was dispatched to my grandparents' address. It was the time before cell phones, so I couldn't call and find out what was going on ahead of time. I arrived to find my grandma unconscious with my granddad by her side. Grandma had leukemia and often commented on how much trouble they had getting IVs on her, which I never understood because, in my opinion, she had good veins. Odd the things I thought about at inappropriate times. I immediately got to work and thought, *We'll see how easily I can access your veins now! I know exactly which vein I'm going for.* Through years of practicing calming myself down, God came alongside me and allowed me to give her better care than anyone else could've given her because I knew her history, what vein to hit, and loved on her during the transport. She never regained consciousness. I stayed by her side and with family at the hospital for an hour or so before I had to mark back in-service. Mom called me in the middle of the night to tell me she had died. While it was difficult for me to transport Grandma, knowing she wasn't going to come out of that, I had the honor to lovingly care for her.

Years later, my stepdad called me on my cell phone, as I was on duty, and told me he's taking my mom to the ER for chest pain and to meet them there. She was in a shock room, with a room full of staff hustling to take care of her. They were going to transfer her to a heart hospital in Indianapolis. The private service that transferred patients from that hospital wasn't quickly available, so I was allowed to transfer her myself. I took another paramedic along with me. The blockage was nearly complete in the right coronary artery, the widow-maker. We made quick time, but I could see Mom deteriorating and knew we had very little time before she went into cardiac arrest. She would suddenly open her eyes and look at me. I would grab her hand and calmly assure her that she was doing well and that we were taking great care of her. But when she would close her eyes, I was getting intubation equipment ready, defibrillation pads out, and calling the hospital, asking for more pain meds and giving them an update.

We took her straight to the cath lab; they were prepping her while we moved her over and started the procedure before our cot was rolled out of the room. When they found out I was her daughter, they pulled me behind the observation glass to stand with the attending cardiologist. I didn't understand why they took me back there at first. When I realized they were allowing me to watch the procedure, I didn't know how I felt about that. But I didn't know which door would lead me out or where to go if I did, so I stood there calmly watching. In a couple minutes, she went into cardiac arrest. They shocked her and got her back. Just a few minutes later, she coded again, shocked her, and got her back again. I do not recommend anyone ever watching something like that on a loved one. Ever. My knowledge and experience gave me insight into how incredibly close we came to losing her. Had she coded five minutes before she did, we wouldn't have gotten her back. It's only because she coded during the procedure that she is alive today. There was only one other time in my life that I so deeply felt a knowing panic due to the impending death of a loved one, and that was when I was trapped in our crashed car while you were hanging unconscious upside down in your car seat, Matty.

Because I followed my path, leading to my purpose, loved what I did, I was given the huge responsibility and gift of loving my family when they needed it the most and making a difference in their lives. And this is love.

CHAPTER 30

A Time to Hate

A heart at peace gives life to the body, but envy rots the bones.
—Proverbs 14:30 NIV

But I trust in you, Lord; I say, "You are my God." My
times are in your hands; deliver me from the hands
of my enemies, from those who pursue me.
—Psalms 31:14–15 NIV

Come to me, all you who are weary and
burdened, and I will give you rest.
—Matthew 11:28 NIV

I never looked for a scripture that authorized me to hate. I always saw scriptures that encouraged me, commanded me, or gave me promises when hate from others rained down on me. You full well know what living in hate feels like. We tried to understand it and make sense of it. We never could. Unlike your dad, his wife, and those many men at the fire department who enjoyed hurting me with nothing but hate, we are not wired to do those things, so we don't understand. I tried for years to convince myself that I didn't hate your dad and his wife for what they were doing to you to get at me, but I was lying. Maybe I fooled others, but God knew my heart. I hated what they were doing to you, both actively and passively. I hated what it was doing to you. I hated that I couldn't get anyone who could help you believe what was happening. I hated the joy they felt when they hurt you and laughed at my misfortunes in front of and to you. I hated the unfairness of what was happening to me at the fire department by men who had no idea who I was or cared to learn.

Because of what alcohol had done to my dad and my family, I hated it. I still do to this day. It was very difficult for me to go on runs involving too much alcohol. I'll admit, anger would well up in me when drunks would yell at me, call me names, speak derogatively at me, touch me inappropriately, and vomit all over. Those were times I was neither gentle nor kind with my words. I hate alcohol.

I have been a work in progress when it comes to hate. It is one of the things I want to heal from. It has affected me and you long enough. Those I hated for the terrible things they did do not know me. I am more than what happened to me. We often referred to our years of enduring hatred as our prison. We literally envisioned ourselves sitting in a prison cell that we couldn't get out of, no matter how badly we wanted to or tried. After years, I began to feel like maybe God had left us. But that was Satan. I know that God was near because we created our prison cell into a place of joy, laughter, and love. Only God could've done that. He never left us. A heart at peace gives life to the body. I want to live. I want you both to live and not just exist. Life comes from peace. Peace comes with healing. I pray this book is your healing too!

CHAPTER 31

A Time for War

He holds success in store for the upright, he is a shield to
those whose walk is blameless, for he guards the course
of the just and protects the way of his faithful ones.
—Proverbs 2:7–8

Discretion will protect you, and understanding will guard you.
—Proverbs 2:11

If you say, "The Lord is my refuge," and you make the
Most High your dwelling, no harm will overtake you, no
disaster will come near your tent. For he will command his
angels concerning you to guard you in all your ways; they
will lift you up in their hands, so that you will not strike
your foot against a stone. You will tread on the lion and the
cobra; you will trample the great lion and the serpent.
—Psalms 91:9–13

He gives strength to the weary and increases the power of the weak.
—Isaiah 40:29

Submit yourselves, then, to God. Resist the
devil, and he will flee from you.
—James 4:7

Like there is a season for hate and killing, it's hard to imagine an ordained season for war. Instead of imagining waging war, I instead caught hold of scriptures that told me how to protect myself from the war Satan brought to me. To obtain the proper tools to protect myself, I made a conscious effort to do as the scriptures instructed. Very often, I fell short and let Satan have access to me. I would write the war plans out where I would see them often:

- Success: be upright (pleasing, righteous, and equitable).
- Shield: blameless walk.
- Guard my course: be just (behaving according to what is morally right and fair).
- Protect my way: be faithful.
- Protect me: have discretion (discipline, self-control).
- Guard me, my understanding.
- Keep harm and disaster away from me: declare that "the Lord is my refuge."
- Resist the devil: "he will flee from me."

My next book about our accident will cover what I'm about to say in greater detail. Many times during the years, you guys lived with verbal and mental abuse at your dad's. I would cry out *to* God, asking why and then imploring Him to please help us. During one particularly hard time for you, Matt, after Ali moved out of your dad's, I was at a deep low. Empty, discouraged, and feeling hopeless and alone, I was sitting on the swing on the front porch. It was raining, and neither of you were home. I was crying and weary. I was yelling out loud *at* God in anger and frustration, accusing him. I remember yelling the words, "Matt doesn't deserve this! He doesn't even want to be here! He was in heaven and asked, multiple times, to not come back! And you made him come back. For this? To be abused? And suffer? Why?"

God immediately and *firmly* told me, *There is a battle being waged in heaven that you know nothing of. My army of angels is holding Satan back, and what gets through I allow, and I will work for my good!*

I heard God. He spoke directly to me—firmly. I didn't hear Him with my ears. I heard Him inside me, inside my cells—in my being. I hung my head and wept, overwhelmed. Satan kept us so beaten down that we couldn't see beyond the battle we were in, always looking ahead to the next one coming. How small-minded. We weren't forgotten. It was so much bigger than just us. Beyond the beautiful stars we loved to look at, there was a battle. A battle that had been going on for years. An army of angels was holding Satan back—an entire army of angels. What if there wasn't an army holding him back? Satan would quiet our story forever and end our lives. God was firmly telling me that He was in control. He was holding Satan back. What got through wasn't an accident. It wasn't because Satan snuck something by God. It was because God knew what Satan was going to do, and He allowed it because it was part of His plan to work it out for His good. I don't know that we will ever know exactly why God brought you back. You. But He has been working on it for His own good since He brought you back. That is humbling. Even as I am typing this, I am trying to see through tears.

A time for war. Satan is waging it, but the battle is God's. Thank You with everything in me for assigning Your army of angels to battle for Matt and for us.

CHAPTER 32

A Time for Peace

Trust in the Lord with all your heart and lean not
on your own understanding; in all your ways submit
to him, and he will make your paths straight.
—Proverbs 3:5–8

A heart at peace gives life to the body, but envy rots the bones.
—Proverbs 14:30

Peace I leave with you; my peace I give you. I do
not give to you as the world gives. Do not let your
hearts be troubled and do not be afraid.
—John 14:27

Peace for us came from being together. Our proverbial jail cell was small and dark. We hated it; we didn't choose to be there, and we didn't deserve it. The cell door swung open, and we were free to leave when you left your dad's, Matty. We ventured close to the opening and looked outside. Bright, fresh, and new, and we were scared. We backed into the cell and voluntarily stood there, huddled together. We had made our cell ours. We filled it with love and happiness from being together. Laughter filled our cell. But we found it overwhelming outside our cell. It was too much. We had been in our cell for twelve years. I was dumbfounded that I hadn't anticipated life beyond being too much and prepared us for that. At different times, one of us would venture out and take some tentative steps away, then feel panic at the freedom and come back to the other two. It was troubling that we chose to return to what almost did us in. But it was known. We knew our cell; we knew how to love and laugh there. We realized that just because the abuse was over and they couldn't hurt us anymore, we carried it with us. Outside the cell, we could trust again, but could we? What if we let our guard down because we thought we were safe and then got blindsided?

God revealed Himself in our cell with its door flung open, and there we still stood. He flooded me with scripture—promises. When we looked toward God and not the overwhelming vastness beyond the cell, we became confident and less anxious. He met us where we were and stood with us patiently until we were ready to keep walking forward. He kept us walking forward until it was too far for us to turn and go back.

"The healing is in the writing." There's peace in the healing. These words were spoken quietly and lovingly to me, repetitively, until I obeyed. Different intensity, but no less impactful.

CHAPTER 33

Memorable Runs and Experiences

Training burn collapse

After less than a year in the fire service, at nineteen years old, we were invited to do a training burn on an old barn with a neighboring department. It was a warm Saturday morning. I had started with a self-contained breathing apparatus (SCBA) on but had taken it off when I saw that nobody else was wearing it and we were outside the barn. I was backing up the nozzleman on a line that was being directed by a young man in the volunteer department, but I knew to also have a career in the Indianapolis Fire Department. I trusted that he knew what he was doing. We advanced the line to the opening of the barn so we could keep the fire from burning the barn too fast. We advanced a little farther until I was standing directly under the heavy wooden beam of the large opening. All of a sudden, the front part of the barn that we were standing under collapsed. I was unaware of any warning that it was about to happen. The beam landed directly on the top of my helmet, collapsing me to the ground. My left leg crumpled beneath me, landing with my foot almost to my left ear. The beam snapped my head back, causing my helmet to fall backward. When debris landed on me, it covered the helmet that was still strapped and pulled taut against my throat. A pile of debris covered me except for a very small opening where I could see the entire barn had been completely consumed with fire and leaning toward me, collapse imminent. I didn't know if anyone knew where I was, and I was completely trapped and unable to move at all. I began to panic, thinking I didn't want to die there!

The man on the pump panel was actually taking pictures when the building collapsed on us, and he just started snapping one picture after the other. I could see in the pictures a number of men running to the pile I was under, pulling debris off me. There are pictures of me emerging out of the rubble, being pulled to my feet, and us running away. The rest of the barn collapsed right where I was, burning freely, only a couple minutes after I got away. What an incredibly close call! After an ER visit, I went back home with a sprained left ankle, bruised left hip bone, bruised trachea—I was hoarse for a few days—and a bruised forehead. It's a miracle I didn't have a broken neck. My helmet had a crack completely through the center of it, and that began my injury-laden fire service career.

I was fortunate to have a handful of shift partners I just clicked with instantly, who had my back when it wasn't popular and caused them grief for standing up for me. Jolly (his real nickname I gave him) was the one who never, ever hesitated to stand with me and stand up for me no matter what. He saw that what was happening to me was wrong, and he spoke out against it. His years in our department were not easy, but man, did we have the best time! The next three standout runs were with him by my side.

Working with Jolly on his last shift at Greenfield FD (April 22, 2020)

Last picture with Jolly and our partner at the
end of the shift (April 23, 2020)

Driving Jolly on Engine 422 on his last shift (April 22, 2020)

Arterial bleed

M22 was toned out in the middle of the night with a neighboring fire department to assist law enforcement with an injured person. En route, we were told to stage off the scene. We arrived in the area, advised dispatch we were staged, and I slid down in my seat and closed my eyes to get more sleep. Over the years, I'd become a master at getting sleep any chance I could. While most of the guys didn't like going on SWAT calls, I thoroughly appreciated it because that was guaranteed time to sleep and not have my sleep interrupted with tones dropping. We stood off the scene for thirty minutes or so before being called in. We found a man in his forties kneeling on the driveway, leaning on the front grill of his vehicle. As this was in the country, it was pitch black out. A police officer told us that a female had cut the patient, and he was bleeding badly. He held his flashlight over the patient so we could see. The cut was deep and long, over his right wrist. His radial artery had been cut, and significant blood had been lost. The patient was holding pressure on his own wound and was conscious but clearly showing signs of shock from the blood lost. I was thankful Jolly was with me, as he always knew what I was doing and where I was headed, and he anticipated, with equipment ready before I needed it. I quickly took the pressure away from the patient, removed the makeshift bandage, and saw what we were dealing with. I placed QuikClot on the wound, absorbent pads, and a pressure dressing that stopped the bleeding.

We began transport to the Indianapolis hospital specializing in hand injuries. During transport, the patient told me his stepdaughter, who had mental health issues, had called and said she was going to commit suicide. He drove to her house, and when he got there, she met him with a large knife in her hands. He talked with her for a while before she lunged at him, swinging the knife as he approached her to take the knife from her. He told me he wasn't mad with her and said that he didn't think she had meant to hurt him, and he was shocked when he started bleeding. She was a nurse and immediately put pressure on the wound to stop the spurting blood. She ran off as law enforcement arrived, leaving the patient to put pressure on it himself. Had she not done that for him, he would've bled out before we got there. As I was giving a report at the patient's bedside, the doctor began to release my pressure dressing. I advised him he would want to stand off to the side so he didn't get hit with blood. I walked out to begin writing my report. The doctor thanked me after he walked out of the room and said the blood spurted across the room and hit the wall. He complimented me on the excellent dressing I placed, saying not a drop of blood had soaked in. He said that not only was the artery completely severed, but so were all the tendons, ligaments, and muscles, and he was already headed to surgery. I don't know what the long-term outcome for the patient was, but he played drums and taught drum lessons for a living. I'm pretty certain that even with excellent care and therapy, he was going to need to make accommodations and adjustments when he could play again.

OD *death*

M22 was called for an unconscious person in our district. It was a busy weekday afternoon. When we walked in, we were met by a teenage girl who told us she couldn't wake her mom up. We walked through the living room where her middle school-aged brother sat. She led us to her mom's bathroom, where we found her unresponsive in a bathtub filled with water. She told us her mom was diabetic, and she thinks it's her blood sugar. Before we ever touched the patient, we could see that she was dead and had been dead for a while. We needed to get her out of the bathtub immediately, but we did not want the daughter to watch how awkward it would be getting her out. Moving someone with wet skin and flaccid limbs in a tight bathroom is not a graceful move for two people. Our engine was still ten minutes out, and we couldn't wait for help. While we drained the water out of the tub and cleared the way through the bathroom and into the bedroom, we asked the daughter questions.

We were told that she spoke with her mom before she got on the bus for school, and she was normal, had no complaints, and had not been sick. When she and her brother got home from school, she called out for her mom. When she didn't answer, she found her in the bathroom. She called 911 when she couldn't get her mom to wake up. We asked her to wait outside to flag down our engine in order to get her out. We shut the bedroom door behind her. We successfully, but not gracefully, pulled the mom out of the tub and into the bedroom. My initial assessment before moving her found pinpoint pupils that would indicate a narcotic overdose. We began CPR and asked law enforcement, who arrived before our engine, to get more info from the daughter and to call someone to have them come be with the kids, as this was not going to have a good outcome. After working on her for the amount of time our protocol stated, without any response, we called the coroner to the scene. Oh, how my heart broke for and identified with these kids, who sat in the living room with hopeful looks on their faces as they looked at us when we walked out of that bedroom.

Mueller Autobody fire

It's November 10, 2017: Twenty-four years on the job, and this shift turned into one of the physically longest and most demanding shifts of my career. Jolly and I were on M22 and were at Station 1 for supplies, parked on the front apron. The tones dropped to investigate smoke or fire. As we were jogging to the ambulance, Jolly was saying he didn't think he would put his gear on until we got there. I was saying that we need to because you never know. Someone was running up to the apron, yelling to get our attention. I turned to tell him he needed to move, as we had trucks that would pull out. When I saw the biggest, blackest column of smoke I had ever seen (we saw TV footage later that showed the column of smoke could be seen past the Indianapolis Airport, miles away), I yelled at Jolly to look! Our eyes were huge!

We headed to the scene ahead of the other trucks and staged at the park across the street, next to the hydrant that would be tagged. M21 pulled alongside us, and as we packed up, I recall telling the other three to be safe, and if ever we lost a firefighter, it could be today. The auto body shop on fire was situated right next to a little strip mall, which would eventually either be smoke- and/or fire-damaged. We pulled a yard lay off E21 and pulled it down a long alley to access the rear of the building. After attaching a second handline to the yard lay, I manned one line in the overhead doorway. The second handline began putting water on the roof, and Jolly pulled a third line that was placed at the eastern edge of the fire building to protect the house sitting just a few feet from the shop. I began on air but ran out. There was neither time nor resources to relieve me, so I could switch out bottles. I disconnected my regulator and just breathed through my mask. For a few minutes, I had help on my line, but there were so few of us that we each manned our own hose lines.

It was a city-recognized holiday (Veteran's Day), so admin wasn't in the office, and we were running minimum staffing with at least one guy in on overtime. Our initial response had two engines, two ambulances, and one ladder truck—ten people total. With one battalion chief running command, three people committed to running trucks, and one climbing the ladder, that left five of us to man hose lines by ourselves. As admin began to show up and mutual aid arrived, nobody came to the rear to relieve us or send us to rehab. I stood in the entrance to that rear overhead door, manning a hose line with incredible pressure being sent to it, not really making a dent in the fire that totally consumed the shop. Cars that were inside the shop had gas tanks adding to the fire load, tires popping, pressurized tanks blowing, a natural gas line that was like a loud blowtorch continually going off, and power line transformers above me blowing and raining sparks down on me for over an hour. I held onto that hose line, my entire body shaking so badly from muscle fatigue that I could hardly stand and hold the line, until I couldn't physically stand any more. *Where is our help?* I wondered.

As I set the line down and walked around to the front, thinking we didn't have any help since the only people in the back were still the original crew, I was shocked to see additional trucks and a staging and rehab area filled with firefighters ready to go to work! I walked up to the rehab/accountability officer and told him the crew in the back was the original crew and we needed to be relieved and sent to rehab. Keeping with years of not caring what I had to say, with disgust and irritation, he told me if I thought I needed a break to just take one. I again reiterated that the rest of the crew needed to be relieved. He dismissed me with the statement that they were big boys and if they needed a break, they could take it. I walked back around to the rear and told every one of them that had been on hose lines from the beginning that there were no less than twenty firefighters standing in staging waiting to get to work and to go to rehab. They were as much in disbelief as I was that no one had come around to the rear. We were on that scene until well past midnight. By the time we got the trucks back in service, showered, and ate something, we only got a couple hours of sleep.

We were still a combined fire department and had volunteers and some part-time firefighters. One of the volunteers who had been on the department for around twelve years or so and always did his hours on our shift was Scott Compton. He responded from home on the dispatch that started the second and third alarms and worked alongside us until all trucks were back in service. He went home to bed and then up to do yard work the next day. His fiancée found him down in the backyard in the early hours of the day following the fire. When the on-duty crew arrived, they found him in cardiac arrest and, unfortunately, were unable to resuscitate him. This was considered a line of duty death as it was within twenty-four hours of the fire. The days to follow were, well, I don't have words. We couldn't believe Scott was gone. Scott was my age; he had just had a birthday three days before. Scott literally never had a bad attitude, was quiet, and always had a positive outlook. I had many a conversation with Scott that were encouraging and uplifting. He respected me and treated me kindly. The world and the people we served were absolutely blessed and better for having this selfless, sacrificial man serve as he did. I am honored to have served alongside Scott.

Mueller Autobody fire. I've circled myself. (November 11, 2017)

Carrying a folded American flag at the cemetery during
Scott Compton's funeral (November 17, 2017)

Shanksville

On October 22, 2012, Hurricane Sandy hit New York. I deployed with INTF-1. For us, it was a no-notice event, as we were at the bottom of the national rotation model and didn't expect to go anywhere. The rotation model rotated all federal teams, so all got a chance to go, even if they weren't closest to the disaster. But the model changed depending on where the disaster was. First, if the state where the disaster happened or was projected to happen had a federal team, that team was pulled off the national roster and held as a state asset. Second, the two or three closest teams, no matter where they were on the model, were sent (because they could get there quicker). That moved all the other teams up.

Because Sandy actually started as a superstorm and wasn't expected to intensify that much, teams along the entire eastern seaboard were pulled out of the model and held as state assets. So when it intensified into hurricane strength and there was great potential for damage and loss of life, we were one of the closest teams. We were sent to Long Island, and with other teams, we did walking searches, walking twelve miles or so each day for days. When we were demobilized, our leaders wanted to take us to the September 11 memorials to remember and honor those who sacrificed their lives. We were unable to do this because of flooding in NYC, so we headed home. They were, however, able to have park rangers arrange for us to visit the Shanksville Memorial in Pennsylvania. I was humbled and honored to walk through the memorial with my team, which numbered around eighty-five members. It's not talked about or mentioned as often as the memorial in NYC, but it is worth going to.

Burning coon and handguns

Yep! You read that correctly! Of course, this is something that would only happen to Haggard! Early in my career, our tanker was dispatched to assist a neighboring volunteer department with a working barn fire. Two of us went, and I assisted in fighting the fire while my partner shuttled and dumped water. The barn sat back off the country road a couple hundred feet. I was backing up the nozzle-man on a handline at the front of the structure. We were standing on the far edge of a concrete pad. A fire scene is loud, and I heard a noise directly behind me, maybe ten feet, that I swore sounded like gunfire. My head jerked around to the right. I couldn't see anything, and I turned my head back when I heard the loud pop again and confirmed, indeed, that is gunfire. I smacked the shoulder of the guy on the nozzle and yelled at him to shut the line down. I whipped around to see a raccoon that had clearly been on fire, smoldering, running around erratically on the concrete pad, and an old (around his eighties) firefighter pointing a handgun at the coon. He popped off another shot as I was turning around, and I about pooped my pants! This was not a large concrete pad—maybe 30 x 30.

I started yelling at the guy, waving my hands frantically in the air, and telling him to stop! He looked at me like I had three heads. I yelled to stop again, and he asked me why, like I'm an idiot. I said, "What are you doing?"

"Trying to put him out of his misery!"

I told him to stop or he'd shoot one of us. The coon is still running around. He shook his head and walked back to his truck. Thinking we were safe now, the firefighter and I moved the line off the concrete pad to the side of the barn, and as we're getting set to open the line back up, I looked over and saw him walking back from his truck with a shotgun! I think, *I'm going to die! I would be the one that would get shot in the head when a bullet ricochets off the concrete and hits me instead of the coon.* We dropped the line and moved to the back of the barn to see what we could help with. I wasn't going to die because of an idiot and his focus on relieving the coon's misery.

CHAPTER 34

Additional Scripture

I wanted to add scripture verses that weren't included in the story but was given to us through prayer.

> The Lord is close to the brokenhearted and saves those who are crushed in spirit. (Psalms 34:18)

He doesn't say that if we believe in Him and walk on the path He's marked out for us, we won't be hurt or crushed. Life isn't easy, and at times, it isn't kind. We've been brokenhearted and had our spirits broken. Our Lord, who sat in our jail cell with us for years, stayed close to us. Not for a few minutes, but for years. We know this promise to be true. When we have periods of doubt, we need to remember what He has done in the past and know that His words are true.

> Be joyful in hope, patient in affliction, faithful in prayer. (Romans 12:12)

Joyful in hope: There are lots of times waiting attached to hope. Those who hope to get pregnant someday may have to wait; hoping the biopsy results are negative requires waiting for days. We are instructed to be joyful in the wait—in hope. And when joy is not a natural response, it's important to recall scripture or dig into His word to see what God is asking us to do.

Patient in affliction: *Affliction* is defined as pain or suffering. We are instructed to be patient in pain and suffering. Oof…cringe! We know my patience level. I wonder, since I don't naturally have a deep well of patience, if instructing myself to be patient repetitively would help. To me, this is a tough one. I know this about myself, so I can pray for God to come alongside me and help me.

Faithful in prayer. Prayer brings us into God's presence. Remember, when we present our requests to God, we start with thanksgiving in our hearts. Maybe we've been patient and faithfully praying, but we feel our hope is waning. Have we offered up thanksgiving and given God our thanks? I know I've left that part out before. Maybe a lot of "befores." Maybe we should write this out and reference it before we pray. Check ourselves. I know I need to.

Final pictures

Medical specialist training in NYC (May 1, 2016)

Ali's Halloween costume was dressing as me (October 31, 2016)

Ali came to the station while I was on duty so I could help
her get ready for her senior prom (April 25, 2017)

Strachan, Chuck, and Lucy (their nickname for me). Together, nearly every deployment of my task force career.

Hurricane Michael (October 13, 2018)

Matt and I on the training ground. He rode out
with me that day. (February 18, 2017)

Moving Ali to Virginia. Matt followed a
couple weeks later. (June 26, 2020)

Drop-off at airport—masks remind us it's during
COVID-19. (November 1, 2020)

Matt and Ali at home to help me move from our
Independence Road house (January 12, 2021)

Us! A perfect depiction—always laughing! (November 19, 2021)

Retirement party from Greenfield Fire (April 6, 2022)

CHAPTER 35

Bel Bel

In August 2022, I began to think it would be helpful and nice if we had a cat (or two or three) to mouse around the house. I prayed about this since it would need to be agreed upon by all four of us who lived here. I brought it up, we all agreed it would help, and I once again prayed for the right cats for us. This may seem trivial to some, but I absolutely *love* cats, and they are not the favorites of the men in the house; therefore, getting the right ones for us was paramount. God directed me to people I knew who had six-month-old kittens they would be willing to part with. These were siblings to quite a few found at the track that I knew were excellent mousers and about 75 percent Manx, so they had bobtails and traits like a dog. After settling on a gray marble kitten and a black one, both girls, I brought them home the first of September.

The gray one we named Mar-Bel (shortened to Bel Bel), and the black one, Jinx (or Jinxy). I quickly and completely fell in love with them, with Bel Bel being the leader of the two. They became excellent mousers, wormed their way into the hearts of the men, and gave the four of us such complete joy, providing laughter as they wrestled and pounced on each other and such sweet moments as they so often slept together. Bel Bel, the more outgoing, became my best friend, following me everywhere, running to greet me every time I came home, loving on me. She loved on us all, but it was obvious that she was my baby girl. I poured my heart out to her and shed many a tear on her. I often thought she was the physical embodiment of God's unconditional love for me. I don't know how many times I will say that they brought us all so much joy. They became outdoor/indoor cats, and we microchipped them so it would activate a door into the garage where they could get out of the elements and away from wildlife dangers. Now, before I go any further about the girls, I'm going to take us through some struggles that we all had to face after I returned from writing my story.

It's December 10, 2023, and I am writing this in my office with Jinxy lying in her box behind me. Our house is decorated for Christmas, our favorite time of year. We are preparing for you guys to come home not just for the holiday but to celebrate my wedding. I believed my story was complete and have been working on final editing and adding the pictures so that I can get this to the printer and have it bound to give to you for Christmas. I've been avoiding this chapter as it is painful for me to write. So, once again, I come before You, Lord, and ask that You write this chapter. Help me to be the fingers on the keyboard, but the message and the words come from You. Guard my mind as I go back and recount the very moment my heart shattered into a million pieces and I lost pure, unconditional love. The healing is in the writing, and I need to continue to heal.

When I went to the ocean on September 10–17, 2023, for solitude to write my story, I fully submitted myself to this God-directed project. My full submission was evident as I typed 95 percent of my story in approximately 45 hours over 4 days, a feat I couldn't have accomplished of my own accord. I knew I would be infuriating Satan. I could sense his fury at my obedience and submission. I discussed with you both how we needed to be in God's presence now more than ever, and I began to specifically pray for protection over each member of our family, including our four cats. Rick Warren states, "Your greatest ministry will come out of your deepest pain." My deepest pain came after my trip. The following is scripture and quotes that were given to me either ahead of or in the midst of our troubles, along with the timeline. God's perfect encouragement at the perfect time. As I remind myself and you every day, God is not surprised by what Satan does. What gets through, He allows, and He *will* work for His good. Period.

Thursday, October 5, 2023:

At 6:00 a.m., you were awakened by the sound of running water hitting the tile in Ali's bathroom. When an incorrectly installed toilet in the apartment above yours caused catastrophic flooding in your apartment, sheer panic and adrenaline kept you going through the move that you had no choice but to undertake over the next five days. With no one to help you, you saved all your property and got every piece of it moved to your new apartment, working feverishly twenty hours a day. I sensed this was Satan's doing. I had been praying specifically for protection over us, but not our property. I was frustrated with myself for not praying for total protection. I expanded my prayers, and the following quotes were given to me:

October 7: "God is using your present circumstances to make you more useful for later roles in His unfolding story" (Louie Giglio).

October 11: "Pray without ceasing because Satan is preying without ceasing" (Toby Mac).

October 12: "Do not let anybody take you back to the place you prayed your way out of" (Toby Mac).

As you continued to have problems from the flood and the move, I continued to pray and encourage you with perfectly timed messages:

October 14: "Your over thinking and worrying about it is assuming that God doesn't know what He is doing" (Toby Mac).

October 22: "God will open doors for you that no man, no hater, and no devil can shut" (Toby Mac).

October 25: "It's hard to hear God's voice when you've already decided what you want Him to say" (Toby Mac).

October 25: If you don't fill your mind with the word of God, the enemy will fill it with fear, anxiety, stress, worry, and temptation.

Sunday, October 29, 2023: Things began to even out as you settled into your new apartment, even though you've had a number of issues with it. After church, Eric and I had a disagreement over an issue that we hadn't been seeing eye to eye on. Both of us said things that weren't mean but were hard to hear. While we talked through it and had a good day, it was still hanging over us as a basic, unresolved topic. After dinner, Bel Bel was sitting on the stairs, where she was ready to go up for me to feed them their dinner. It had been such a beautiful day that I decided to feed them in the garage and leave them outside overnight, as I knew there wouldn't be too many more nice days before the cold moved in. I called out, "Good night, girls! I love you!" Then I went inside to get ready for bed and the busy work week ahead.

Monday, October 30, 2023:

> Be alert and of sober mind. Your enemy the devil prowls around like a roaring lion looking for someone to devour.
> —1 Peter 5:8

> The thief comes only to steal and kill and destroy.
> —John 10:10

My alarm went off at 4:40 a.m., and I was up to begin my day. The girls knew my routine and that I was up either because they could hear me or saw lights coming on. My morning routine didn't vary, but this particular day, I had a couple extra minutes before heading out the door. I got on my phone to check emails when I felt a very strong prompting to go outside early. As I turned the laundry room light on to get my coat on, gather my things, and shut the alarm off, I knew they would be right outside the door waiting on me—Bel Bel at the door and Jinxy a few feet behind. As I walked out and turned to lock the door, I was surprised they weren't there, so I began to call out, "Where are my girls? Where are my Bel Bel, Jinxy?" I heard a very faint and short cry out to my right. I stopped and cocked my head, unable to see but trying to locate the sound. I called out and again heard the faint cry. I knew it was an injured cat but thought it was a stray or a neighbor's cat. I unlocked the garage door and dropped my things in my car so I could go out and find the cat. As it was still dark outside, I used the light on my phone, and with another weak cry, I found Bel Bel lying on her right side in the grass at the edge of the front walk. Her eyes were open, and she was looking at me. I initially thought she had eaten a mouse that was poisoned, and she was sick. But as I shone the light fully on her, I saw the most horrible sight that I will never forget.

Even now, as I see it in my mind's eye, it makes me nauseated, my breathing quickens and becomes shallow, and I cry. My sweet baby girl, my best friend, my Bel Bel, had an injury to her abdomen that was so devastating I knew instantly she wouldn't survive. As I cried out, "No, no, no, no, no!" I ran inside to wake Eric up for help. As I ran back outside, I was greeted with another weak cry. I kneeled down and began to love on her, crying and praying while Eric got a towel to wrap her in and place her in a box. I could see in her eyes that she was fading. As we moved her into the garage and made phone calls to call off work and figure out where to take her, I cried and loved on my girl. We saw Jinxy walk by cautiously and hesitantly. We didn't see her again before we left with her sister, her protector, and her constant companion.

The drive to the emergency animal hospital took around forty minutes, encountering suddenly slowed traffic from a very recent crash in a construction zone that law enforcement had moved to the side of the interstate. I was loving on my girl, sobbing, and praying nonstop. When we slowed for this traffic, I began to cry harder and beg God to please clear the way, which He did. Twice, Bel Bel cried an agonizing and long cry and attempted to get up with her front legs. We realized that the injury included severing her spine, and she couldn't move the bottom half of her body. I began to pray that God would take her quickly and that she wouldn't suffer anymore. We loved on her; she calmed, and I began to praise God. As I sobbed, I cried, "I will praise You in this storm!" I thanked Him that I came out of the house when I did; that Bel Bel could call for me; that I heard her and found her quickly; that Eric was home that morning; for jobs that we could call off late; that we worked well together; that He cleared the way for us; that He gave the girls as a gift; for the love we shared; that the weather was clear that day; that I wasn't out of town. When we arrived, they quickly took her to the back to assess and establish an IV on her. I collapsed to the floor, sobbing uncontrollably, when they wouldn't let us go back with her. I did not want her to suffer anymore, but I did not want her to die without us by her side. I began to pray that they would get the IV and that she would hang on until we could be with her.

The tension and the distance that had been between Eric and me the day before were completely gone. We grieved together and supported each other. Eric was the rock who took care of everything so that I could be fully and completely present for my best friend, my baby girl. This was a gift he gave to Bel Bel and me. God answered our prayer, and we were brought back to be with her. There was nothing they could do to save her, and we loved on her while they relieved her of her pain. My pride, my stubborn independence, and my unwillingness to let anyone—even my fiancé and God—be close to me died with Bel Bel that day. From that moment on, I changed; though I continue to grieve, I feel a peace and calm I have never felt before. I said countless times a day, still, "I will praise You in this storm." Casting Crowns has a song called "Praise You in This Storm." These words resonate with me: "As the thunder rolls, I barely hear you whisper through the rain, 'I'm with you.' And as your mercy falls, I raise my hands and praise the God who gives and takes away. I'll praise you in this storm, and I will lift my hands, for you are who you are, no matter where I am. Every tear I've cried, you hold in your hand; you never left my side, and though my heart is torn, I will praise you in this storm."

We came home with Bel Bel so my parents could say goodbye. Jinxy was in the garage, and I brought her in. I spent every free moment I had with her in the weeks to come. We buried our girl where we found her and put up a beautiful memorial stone with the rocks that mark her grave. I know she left her spirit in Jinxy and me because she knew we would struggle the most and need each other. I felt her spirit as Jinxy exhibited four different behaviors that only Bel Bel would do, exactly the way she used to. Jinxy used to be so skittish that she wouldn't hardly let us pet her. Now, she comes to us each day for love and reassurance, follows me around, and lies beside me, as she is right now. I spent hours praising and praying. I still get nervous and anxious when I walk outside that door when it's dark outside; my mind immediately replaying her weak cry and seeing her wounded. I know it was Satan who came to steal and kill our joy, but he did not destroy my relationship with Eric. Through our grief and my praising God in the midst of the storm, our relationship became cemented and solid. God answered my cries of heartbreak with scripture and encouragement.

October 31:

> I will say of the Lord, "He is my refuge and
> my fortress, my God, in whom I trust.
> —Psalms 91:2 NIV

All the forces of darkness cannot stop what God has ordained.
—Isaiah 14:27

November 2: Spending a couple hours with Jinxy every afternoon helped us both. She has a habit of lying her teeth on our hands when she doesn't want to be pet anymore, but she rarely breaks skin. Today, she did that to me, but then she clamped down. The puncture wound became infected, and I ended up in the ER and almost needed to be admitted for treatment of the infection that was spreading quickly and making me terribly sick. My encouragement: "God doesn't stop the bad things from happening; that's never been part of the promise. The promise is: I am with you. I am with you now until the end of time" (Madeleine L'Engle). I began praying very specifically: "God, please place Your mighty hand over each member of my family and our cats. Shelter us under Your wing and place us against Your body so we can rest and feel Your strength. Please guard, shield, and protect each of us and our property. Banish the enemy from us." I prayed this very specific prayer many times each day. He answered me not only with the peace I carried along with the grief, but He sent verses to me day after day:

November 4: "Hardship often prepares an ordinary person for an extraordinary destiny" (C. S. Lewis).

November 8: "He lifted me out of the pit of despair, out of the mud and the mire. He set my feet on solid ground and steadied me as I walked along" (Psalms 40:2 NLT).

November 9: "The Lord himself watches over you! The Lord stands beside you as your protective shade" (Psalms 121:5 NLT).

November 10: "Those who live in the shelter of the Most High will find rest in the shadow of the Almighty." (Psalms 91:1 NLT).

November 11: "He will cover you with his feathers. He will shelter you with his wings. His faithful promises are your armor and protection" (Psalms 91:4 NLT).

November 12: "This is my command—be strong and courageous! Do not be afraid or discouraged. For the Lord your God is with you wherever you go" (Joshua 1:9 NLT).

November 16: "We ask God to change our situation not knowing He put us in the situation to change us" (Toby Mac).

November 26: "The Lord is my strength and shield. I trust him with all my heart. He helps me, and my heart is filled with joy. I burst out in songs of thanksgiving" (Psalms 28:7 NLT).

We believe coyotes came right up to the door that I was literally about to come out of. Jinxy escaped into the garage, and Bel Bel ran toward a tree. She was grabbed, and the coyote swiped at her and was in the process of carrying her off when I came out. She was most likely dropped as they ran off. Thank you, Jesus, for the prompting to go out early, and I immediately followed that prompting. I praise you, Lord, that we could wrap her with our love as she died, and she wasn't carried off for us to search endlessly to never find her and die alone.

Wednesday, November 29, 2023: As problems continued in your new apartment, you both decided you wanted to move home before the planned time. After discussing it further, you decided to start hunting for jobs in Indiana. We began to pray about it, and you started filling out job applications. As always, daily, we are given emails, encouraging statements, and scripture to help bolster our faith and trust in His provision and timing.

- "Not sure who needs this, but one day you're going to look back and be grateful that life played out God's way and not yours" (Toby Mac).
- "In the same way, I will not cause pain without allowing something new to be born" (Isaiah 66:9 ERV).

When I lay my head down at night, I begin to deeply miss my girl and can hear her cry for me. I am a very visual person. I have always used the practice of visualization to bring something to life or make it more real for me. This is the time when my mind is most vulnerable to Satan's attacks. I could very easily begin accusing God, blaming myself, and reliving that horrific morning. However, from that first night and every night since, as I lie there with my eyes closed, I take a deep breath, and I see myself bend down to pick Jinxy up, and I stand in front of God, our Father, who is sitting on His throne in heaven. I say to Him, *Oh, Father, I am tired and need sleep. May I crawl into Your lap, curl up, and rest? You created Jinxy and I. We're broken, and only the one who created us can heal us.* I then envision crawling up, setting Jinxy on His lap, lying back, and then I feel my body fully relaxing, all tension gone. I then call Bel Bel to us. Every single night, I've then fallen deeply asleep as Bel Bel curls up on my right side with Jinxy curled up with her.

Satan came after us because I'm doing what God told me to do. God let it through, and He *will* work it for His good. He very clearly told me that, and I have no reason not to believe Him. Satan does not want my story told. Bel Bel was created for us—for me. But only for a very short time. During my time with Bel Bel, I knew in flesh what God's love offers me every single day, in big and small ways. I will honor her life and memory by telling her story too.

My Bel Bel

Bel Bel with Jinxy

CHAPTER 36

Putting It All Together

One final run of scripture, some of it repeated from earlier in the book. In the summer of 2022, I was working a shift at the races, exhausted, working nonstop at the horse track, body not where I wanted it to be, left foot hurting from the torn plantar tendon and plantar fasciitis, missing the fire department and task force, loving my job but feeling like there's something more I'm supposed to be doing. I was going through my phone looking at pictures of verses and sayings (from K-Love, Toby Mac, and others) I had saved, reading through them to encourage me. I began to write them down one after another, circling and underlining key words, seeing a pattern of words: beginning, ending, trust, better, purpose, storm, and still or quietly. As you read the compilation, one right after the other, imagine my exchange with God. God wanted me to know: This isn't an ending, it's a beginning. Sit quietly and be still in the storm. I have a purpose for you that is better than you can imagine. Trust me. Read this again. Absorb it, and let it sink in. My babies, this is for you too.

The compilation

If you could see what's coming, you wouldn't stress so much about what's happening. When God gives you a new beginning, it often starts with an ending. If God shuts a door, stop trying to open it. Trust that whatever is behind you was not meant for you. God says,

Stay patient. My timing is perfect. I have something bigger planned for you, and trust me. You're going to love it. There is a powerful moment when you exhale and let go of what once was and inhale and embrace the wonder of what will be. Sometimes the bad things that happen in our lives put us directly on the path to the best things that will ever happen to us. This is about two choices life gives every person: either sit, sulk, and dwell on how unfair life is to you, or you could try and figure out how to make the bad situation work in your favor.

Maybe it's not working out because God is working out something better. One day, you will tell your story of how you've overcome what you're going through now. And it will become part of someone else's survival guide. Sometimes God doesn't give you what you think you want, not because you don't deserve it, but because you deserve better.

God will give you what you need when it's time. Allow this hard season to grow you. God wouldn't have allowed it unless He had a purpose. Sometimes God's blessings are not in what He gives but in what He takes away. You don't have to understand the plan to trust that God has a purpose. Sometimes you just have to find the courage to let go of things you cannot change.

Sometimes God's best way to add to your life is to subtract from it. Certain things have to end so better things can begin. Every loss isn't a loss. Faith tells me that no matter what lies ahead of me, God is already there. Focus on me. Not the storm.

Don't let go of my faith because of what I have yet to see. Let all that I am wait quietly before God, for my hope is in Him. Letting go doesn't mean giving up, but rather accepting that there are things that cannot be. "Don't let your hearts be troubled. Trust in God and trust also in me" (John 14:1 NLT). The process He is taking you through has a purpose. Trust God, even when it does not make sense.

Letting go doesn't mean giving up, but rather accepting that there are things that cannot be. God is telling me to be still. Don't stress over what is going on. Stop figuring it out with my own understanding. God is saying to leave it for me to handle. "But those who trust in the Lord will find new strength. They will soar high on wings like eagles. They will run and not grow weary. They will walk and not faint" (Isaiah 40:31 NLT).

God is not in a hurry. You are. It is why you are tired. It is why you are anxious, stressed, and disappointed. Surrender your timeline in favor of His peace. Sometimes our lives have to be completely shaken up, changed, and rearranged to relocate us to the place we are meant to be. "The Lord will work out his plans for my life—for your faithful love, O Lord, endures forever" (Psalms 138:8 NLT).

Maybe it's not working out because God is working out something better. Sometimes it takes your breakdown to create your breakthrough. Sometimes our greatest blessings come from our biggest disappointments. You see what I have carried you through? Wait until you see where I carry you. God brought you to this moment on purpose. Remember that. On purpose. To prepare you for your future.

"When you go through deep waters, I will be with you. When you go through rivers of difficulty, you will not drown. When you walk through the fire of oppression, you will not be burned up; the flames will not consume you" (Isaiah 43:2 NLT). One reason people resist change is because they focus on what they have to give up instead of what they have to gain. Don't worry that you're not strong enough to begin. It is in the journey that God makes you strong. "He gives power to the weak and strength to the powerless." (Isaiah 40:29 NLT). When you hold on to your history, you do it at the expense of your destiny. Not all storms come to disrupt your life. Some come to clear your path. If your path is difficult, it is sometimes because your calling is higher.

God wouldn't have allowed it unless he had a purpose. There are times when God asks nothing of His children except silence, patience, and tears. "Be still in the presence of the Lord, and wait patiently for him to act. Don't worry about evil people who prosper or fret about their wicked schemes" (Psalms 37:7 NLT). Sometimes you are delayed where you are because God knows there is a storm where you are headed. "The Lord will work out his plans for my life—for your faithful love, O Lord, endures forever. Don't abandon me, for you made me" (Psalms 138:8 NLT).

"The Lord helps the fallen and lifts those bent beneath their loads" (Psalms 145:14 NLT). I had to make you uncomfortable; otherwise, you would have never moved. If you could see what's coming, you wouldn't stress so much about what's happening. Don't underestimate what God is doing in your season of waiting.

So, make friends with whatever is next. Embrace it. Accept it. Don't resist it. Change is not only a part of life; change is a necessary part of God's strategy. To use us to change the world, He alters our assignments. God is writing my story.

So much was given to us by God, through scripture, that He truly did write my story through the years.

Leaving a legacy means giving something that will be valued and treasured by those who survive after my death. It requires thought to ensure that any items that have meaning to me will also have meaning to those I designate to inherit them.

This is my legacy to you. This is my story.

ABOUT THE AUTHOR

After thirty-three years in the fire service, Beth Frenzel continues to work as a paramedic for special events. She resides in rural Indiana with her family. Beth enjoys being outdoors, traveling, and taking long walks. She can be found taking pictures of sunsets and sunrises.

www.ingramcontent.com/pod-product-compliance
Lightning Source LLC
LaVergne TN
LVHW040937131224
799041LV00012B/166